Bonsai
the Blue Sky way

Bonsai
the Blue Sky way

A simple guide to Bonsai Horticulture

Dave Seymour

BLUE SKY BONSAI

Bonsai, the Blue Sky way: A simple guide to Bonsai Horticulture

Author: Dave Seymour

First published in 2024 by Blue Sky Bonsai

Copyright © 2024 by Dave Seymour, Blue Sky Bonsai

All rights reserved. No part of this work may be reproduced or transmitted in any form, in any language, without the prior written consent of the copyright owner. All word marks within this publication are owned and copyrighted by the author, Dave Seymour. All photographs and drawings are owned and copyrighted by the author with the exception of the front cover image and title page image. Front cover image and title page image thanks to Sage Ross, October 2008: Trident Maple (Acer buergerianum), National Bonsai & Penjing Museum at the United States National Arboretum. Image derived from original photograph by Sage Ross, multi-license with GFDL and Creative Commons CC-BY-SA-2.5 and older versions (2.0 and 1.0).

Legal disclaimer: The information provided in this book is intended for educational purposes only. While every effort has been made to ensure the accuracy of the information, the author and publisher assume no responsibility for any errors, inaccuracies or omissions. The author and publisher disclaim all liability for any loss, injury, or damage incurred as a result, direct or indirect, of the use or misuse of the information provided in this book. By reading this book, you acknowledge and agree to assume full responsibility for your actions and their consequences.

ISBN: 978-1-0686985-0-7 (hardcover)
ISBN: 978-1-0686985-1-4 (paperback)

blueskybonsai.com

Contents

Preface 7

Part 1—Essentials 9

Golden Rules 10
    Why you're here 11
    Doing bonsai: practical advice 14
    Bonsai Beginners—Start here! 15
    Cultivating bonsai trees 17
    Indoor or outdoor bonsai 25
    Guidance by climate 28
    Tree talk for bonsai lovers 31

Part 2—Bonsai Reference A-Z 39
    Bonsai benches / bonsai garden 40
    Bonsai sizes 45
    Bonsai styles and styling 50
    Bonsai tools 62
    Branch pruning: why, when, how 65
    Creating an apex 80
    Defoliation 81
    Developing a bonsai 83
    Fertilizer science 85
    Fertilizers—recommendations 98
    Flowers and fruit 102
    Front of tree & planting angle 108
    Moss 110
    Nebari—the surface roots 115
    Pot choice 118

Pot refresh 122
Propagating trees 123
Repotting: why and when 129
Repotting: how to repot a bonsai 142
Repotting FAQs 146
Reviving a dying bonsai 154
Slip potting 160
Soil 161
Tourniquet 169
Trunk chops, trunk thickness and taper 172
Watering bonsai 176
Watering systems compared 180
Wiring branches 189
Wiring the trunk 196

Advanced techniques 198

Index 199

Acknowledgements and further reading 203

About the author 204

Preface

I love trees; always have done. Especially small ones in pots.

I got my first bonsai in 2006—a Serissa. It died about a year later. My next few bonsai trees all died too. I refused to give up, and decided to deepen my understanding in the whole area of container-based tree growing. It was clear that I had an important gap in my knowledge. In short, I needed to understand bonsai horticulture. So for the next few years I attended workshops, bought and read hundreds of bonsai books, magazines, and articles. And yet, I always found myself missing a go-to book to explain in simple terms how to keep a bonsai healthy, vigorous, or even just alive. I needed an easy-to-use reference guide of bonsai concepts and tasks, told from a tree-health point of view.

I am originally from the United Kingdom, and for the last 25 years I've lived in Madrid, Spain. I'm continually astounded by the stunning Spanish blue sky, and for many years I have found myself taking photos of trees under blue skies. That, combined with my love of bonsai led me to start filming the bonsai work. The educator in me wanted to teach others exactly what I wished I had learnt all those years ago... with some music and visuals that you don't get in a bonsai workshop.

My YouTube channel is called "Blue Sky Bonsai". In my videos, I try to explain the horticultural essentials—information that I wish I had understood before I made all the bonsai mistakes along the way. And now I feel the need to do the same on paper with this book, which I hope will fill that same gap for you, and so you don't make the same mistakes as me. My aim for this book is to be informative, accurate and useful, but also enjoyable to read and simple to use.

You can also use this book as a reference guide to my tutorials. Simply look up your desired subject in the reference A-Z section. To see the relevant video tutorial, use a mobile phone to scan the QR code in the section header. For example, this code (right) is for the Blue Sky Bonsai channel page on YouTube.

Thank you, dear reader, for taking this next step in your bonsai journey. I sincerely hope you enjoy the book, and get as much pleasure from your bonsai trees for many years, as I have done.

Trident Maple "Miyasama", height 45cm.
(Camera lens 17mm at f/4, distance 61cm).

Part 1—Essentials

Golden Rules

1. Don't let the roots get dry.

> Exceptions: succulents

2. Don't prune the roots in autumn or early winter.

> Exceptions: tropical species

3. Don't prune off all a conifer's foliage.

> Exceptions: only if you want to kill it

4. Don't leave your pot standing in a pool of water.

> Exceptions: cuttings before roots develop
>
> (root-rot can't happen if there are no roots)

5. The pot must have drainage holes. No exceptions.

Why you're here

This book shows you how to keep your bonsai alive, healthy and happy from the day you get it home, to . . . forever.

After reading this, you might not be a bonsai expert, but you will be able to make your potted tree look truly beautiful, like a small, believable version of a real tree in nature. And most importantly, **you will be able to keep it alive and growing vigorously** for much longer.

Spruce forest

In this book, I'll show you what I've learnt the hard way, so you don't make the same mistakes as me, and so you can learn the easy way.

If you're an absolute beginner, and want a simple "cheat-sheet" of basic bonsai care, you could jump to *Bonsai Beginners—Start here!* on page 15.

But my guess is that before you picked up this book, you probably already had at least a couple of small trees in pots, in various stages of development and sizes. It's also possible you already know that "bonsai" was originally a Japanese term for "planting in a tray", or "tree in a container"; the same in Chinese.

Dawn redwood grove flanked by two Deshojo Japanese maples

This book removes the illusion of mystique from the ancient art, and shows you essential container-based horticulture without mystery or snobbery, so that you can truly understand and enjoy your bonsai.

So, why are you here?

What got you into bonsai? ...

Did someone gift you a small bonsai? Did you already have several houseplants and decided to try some small trees in pots?

Or is it because you get a surge of warmth every time you see a beautiful old tree in nature, and you want a part of that nature at home?

Perhaps it's because you saw Mr. Miyagi pruning several junipers in the movie *Karate Kid* (1984)? Or the amazing bonsai trees in *Dune* (2021)?

Maybe you saw someone's video on YouTube and decided that you must have a go at that wonderful hobby?

There are many different ways that people get into this horticultural pursuit. The love of small trees can become an addiction, and it stays with you for the rest of your life.

Doing bonsai changes the way you look at trees in nature; you can no longer pass by a large tree without admiring the fantastic surface roots (right), trunk movement, branch structure, or bark texture. And maybe you take snaps and upload the photos to Instagram or Flickr. I certainly do.

Superb surface roots

The love of small trees in pots is a very noble pursuit when you are studying, understanding, and propagating them. If your loved ones suggest that you are becoming obsessive over some little trees, tell them that **it is a passion; not an obsession**!

... And what does "bonsai" mean to you?

Is it an artistic representation of the harsh elements of nature acting on the tree, contrasted against the tree's need to grow and reproduce? Or is it just lovely to have a pretty little tree at home?

Is bonsai an artistic display of a tree in a pot? Or is it the art of creating that display?

In modern bonsai culture, we often talk about "doing bonsai" as a verb, to represent the enjoyable work, the skill, the passion and the patience, as much as it is a word for a tree in a pot.

Japanese white pine

While images of beautiful bonsai are ubiquitous, I am convinced most people don't get it; they don't see the decades of work that went into each spectacular tree. They don't realize that a bonsai is never finished. That's why doing bonsai – as a verb – suits this pursuit so well: it's not just to have a piece of art; it is because we enjoy doing it, and have the patience to shape our planting into a beautiful tree over many years.

But you—you already get that; I believe that's why you're here.

Doing bonsai: practical advice

Over the years I have accumulated a few pieces of experience that have made the hobby more enjoyable and rewarding.

Do:

- **Take photos** of each tree, every time you do some work on it. For example, when root pruning, take a photo of the roots before and after, as well as the final image when repotted.

- **Start a photo archive**, either private or online. Make one album or folder per bonsai. This way, you visually track the progress of each tree. I use Flickr.com for this purpose.

- **Choose somewhere** to do your bonsai work that makes it enjoyable and does not feel like a chore.

- **Join a bonsai club** or attend local workshops. It'll make it quicker and easier to make your own decisions on trees.

Don't:

- Don't do a bonsai task if it feels like an arduous chore. Likewise, don't start your bonsai work if you have a "hard stop" deadline. **All bonsai work can wait for another time, except for watering.**

- Don't start bonsai work under low-light conditions or in twilight. Use good lighting to avoid pruning mistakes, unwanted planting angles, or worse, physical injuries.

- Don't bother making videos of your bonsai work, unless you thoroughly enjoy camera work and videography. It is an entirely separate hobby from bonsai, takes a lot of time, and can detract from the enjoyment and tranquility of bonsai work.

Bonsai Beginners—Start here!

Welcome to the fantastic world of beautiful mini-trees in small pots!

So, you're new to bonsai and need to know the basics of bonsai care. Some advice depends on your species of bonsai. That's your first task—find out what species of tree or plant you have. And importantly:

 Find out if your bonsai species should be kept indoors or outdoors, in your climate. Then **place it in the right location**. Outdoor trees slowly die in the house, while tropical bonsai die outside in the cold. See *Indoor or outdoor bonsai*, page 25.

 Never let the soil get completely dry. The majority of bonsai deaths occur shortly after they were not watered. See *Watering Bonsai*, page 176, and *Reviving a dying bonsai*, page 154.

New bonsai, DAY 1 / MONTH 1

Once you've positioned your bonsai, let it sit in its pot for a few weeks to get **accustomed to its new environment**. Water it whenever the soil starts to get dry.

One thing you must practice in bonsai, is patience. Don't prune it yet; don't fertilize it yet; just water it and wait. Some leaves might turn yellow or even drop. This is normal for some tropical species, and can happen while the plant gets acclimatized to its new environment.

New Chinese elm needs work but it can wait

MONTH 2

After a few weeks, gently ease the whole root ball out of the pot and **inspect the roots**. If they look unhealthy, or terribly overcrowded, you could "slip pot" the entire root ball into a bigger pot. That's a quick and simple process, and could save the tree's life. See *Slip potting*, page 160.

Overcrowded roots

But if the roots are not looking overcrowded, place it back in its pot and delay the repotting decision until next Spring. The safest time to repot is Spring when the leaf buds are swelling, ready for bud burst.

Beyond MONTH 2

Your next steps – after the new environment and root check – are dependent on what size the bonsai is, and what size you want it to be.

If you want the tree to grow a thicker trunk before you start refining the branches, then don't prune it; just let it grow out, unhindered.

Did you buy it as a pre-bonsai in a large plastic grow-pot? Or is it already in a bonsai pot? Keep in mind that trees in small bonsai pots almost completely stop thickening, so if you want it to thicken more then you'll need to put it in a bigger, deeper pot for 3 or so years and let it grow wild; as big as possible. If you keep pruning the branches and the roots, you will keep the tree trunk small.

I recommend you buy two or three "first bonsai" trees, and try different plans for each tree. First one, keep in its small bonsai pot, and start refining it as a small tree. Bonsai two, slip pot into a deep pot to thicken it up for a few years. Get an outdoor and an indoor tree. I recommend a Chinese Elm for outdoors and a Ficus Microcarpa for indoors. These are both resilient species and are forgiving of all types of bonsai work.

Cultivating bonsai trees
Art or horticulture?

Arboriculture is the cultivation, management and study of woody plants. It is, essentially, the horticulture of trees. And of shrubs and vines. In this sense, *Blue Sky Bonsai* is all about arboriculture.

Juniper cascade drawing

"Why is this important?", I hear you ask. Here's why:

Before you become an artist in oil paintings, you need to know a bit about the medium of oil paints, quality of materials, canvas surface preparation, how to mix colours, and some known painting techniques.

Likewise, before you become an artist in bonsai, you must know the basics about the medium: live trees. And the more you understand the inner workings of trees and plants, their roots, stems, foliage, species variations and seasonal behaviour, the more easily you will be able to harness your knowledge to cultivate a piece of beautiful living art.

Ultimately that's what separates bonsai from other types of art: you are creating a horticultural marvel that is never finished; continually growing, slowly changing, and always beautiful. Living art.

What kind of tree is a bonsai?

Almost any type of tree or tree-like plant can be made into a bonsai.

There is an occasional misconception that all bonsai trees are dwarf species from Asia. While bonsai are typically small, they are not necessarily dwarf species, and can originate from any part of the world.

However, some of the most popular bonsai species are Asian by origin, such as Chinese elms, Chinese junipers, Japanese maples, and Japanese pines. These species make superb bonsai due to their resilience and beauty, perhaps bolstering the belief that all bonsai trees are from Asia.

Many European, African, American and Australian trees also make excellent bonsai; also species native to the tropics. Succulents too, such as the Portulacaria afra, sometimes called "dwarf jade". This is just one example of how a common name can mislead us: it is not a tree, nor is it a jade, nor is it a dwarf cultivar. But it makes a great bonsai.

> **Note:** Palm trees typically don't make good bonsai, because of their large leaf size and their lack of branching structure.

Terminology: bonsai, bonsais and bonsai trees

In English, the plural of "bonsai" is simply "bonsai"; there's no such word as "bonsais". For clarity, we sometimes use "bonsai trees" to talk more fluidly about multiple trees in pots.

Furthermore, "bonsai" can encompass not only the tree, but the overall planting, including the pot and other elements such as rocks. And, since a bonsai planting can be any woody shrub, when we say "bonsai trees", we should really say "bonsai plantings". However, for simplicity and clarity in this book, I use "bonsai" and "bonsai trees" interchangeably.

Do you want a bonsai or a tree?

Do you want your potted tree to look like a typical bonsai? Or do you want your bonsai to look like a real tree in miniature?

If you search for "bonsai images" in a web browser, you see many pictures of beautifully sculpted, perfectly pruned bonsai in high-quality pots. While gloriously beautiful, and artistically wonderful, most of these trees are both above the price range and outside the development time available to the average bonsai enthusiast.

Beautiful mature juniper

Crab apple with wavy trunk

And so, we adopt this image in our mind of the typical bonsai tree shape; the familiar "S-bend". I bought this crab apple (left) in 2016, and initially I liked the wavy trunk movement. But it has some challenging problems. The top portion of the trunk is nearly as thick as the base. The minimal trunk taper is only made apparent by the top being nearer the base, due to the curvy trunk.

So let me ask you: With this contrived, curvy trunk, does it look like a real tree in nature? It doesn't to me, but maybe your imagination is better than mine. The thing is, fruit trees don't naturally grow like this.

Some nurseries grow their young plants in a way to sell them for the optimum price in the shortest time. Developing a shoot with an S-bend is a quick way of making a young, thin tree trunk appear to have more taper and visual interest in a few quick years. And some people like the appearance of that contrived S-bend. That's why it has become visually synonymous with many people's perception of what a bonsai is.

Genuine trunk taper grows slowly over many years, and when it is developed well, it looks a lot more convincing; it can make a bonsai appear more like a real, old tree in nature.

Trunk taper on a trident maple

Trunk taper – from thick at the base to thin, fine twigginess at the top – is a desired quality in a bonsai tree, to reinforce the impression of a mature, old tree in nature. See *Trunk chops, trunk thickness and taper*, page 172.

Contorted beech in northern Spain

You *can* find some very interesting examples of curvy trunks in nature. This amazing beech (left) grows on an incline in the magical Faedo de Ciñera forest in northern Spain. A tree like this would make a curious addition to a bonsai bench or garden. Its twists tell a story of contorted growth in a tough environment; its surface roots suggest subterranean rocks and soil erosion. The tree is beautiful, and it doesn't look like a "typical" bonsai.

Now let's go one step further: consider a large, mature old deciduous tree in a picturesque field. I bet you have never seen one with a trunk that has curvy S-bends. They tend to look more like this old English oak (left) in Derby, UK.

Here, we can see many branches, and the unmistakable image of a very old tree, growing relentlessly despite the harsh conditions of nature over hundreds of years.

If we look more closely, (right), we see that the enormous, fat old trunk is hollow and tapers rather erratically to a thinner leader, which tapers again further up. Consider also the multiple branches to the right; normally we permit a bonsai to have only one branch emerging at any point or node, but this is a purely aesthetic imposition and does not always reflect the reality of nature.

So, once again I ask you:

Do you want your little tree in a pot to look like a typical bonsai? Or do you want your bonsai to look like a real tree, in miniature?

For me, it's the second: a miniature representation of a real tree in nature. It should have shape, scale and relative proportions that make us believe, for a moment, that it could be a real tree—whether on the edge of a cliff hanging on for dear life, or in the African savannah, or in a peaceful pine wood, or simply in a nearby field.

In reality, most bonsai plantings fit somewhere between the two ends of the scale. But as a goal, it is both fulfilling and artistically pleasing to develop each bonsai gradually but persistently towards the objective of appearing like a real tree.

Trident maple Miyasama

You are the artist; you choose the image.

Leaf size, internode length, proportions

Leaf size is an important factor in convincing the viewer that a bonsai looks like a real tree. A fully grown tree has tens of thousands of leaves which are tiny compared to overall tree size. This is why species with naturally small leaves make great bonsai material. And it is also why one part of bonsai practice is centred around reducing leaf size.

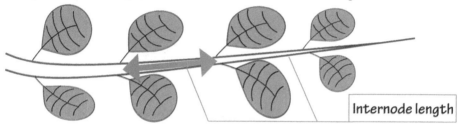

Similarly, the internode length – the distance between leaves or side shoots on a stem – plays an essential part in the realistic portrayal of a tree. Long, bare twigs can break the illusion of maturity, exposing the true young age of the plant. As well as trying to decrease the leaf size, we also attempt to reduce internode lengths, to help develop a more compact and realistically proportioned tree.

There are specific tasks that we perform on certain trees with the sole intent of limiting the extension of internodes while each shoot grows. For example, on Japanese maples following budburst in Spring, we can use tweezers to pinch out each second pair of leaves while they are forming. This reduces the stem's vigour slightly, so that the remaining energy goes to reforming the next pair of leaves, more than to extending the stem.

Second leaf pair emerging on a Japanese maple

However, many tree species with large leaves, such as deciduous oaks and field maples, can also make stunning bonsai. And because of the impossibility of growing thirty-thousand leaves on a 30cm (12") tree, we sometimes view a bonsai as a representative depiction of a tree, rather than an exact, scaled-down version of a full-size tree. For example, one large leaf on a bonsai might represent a whole branch of the tree in your mind's eye.

So, the leaf size is a significant factor, but not critical. What is more contributory to a realistic portrayal of a tree is the shape and size of the trunk in relation to the branches.

Sageretia, 2019

It takes an artist many years of work and patience to cultivate a tree that genuinely looks like a fully grown mature tree in nature. On the other hand, it can take just a few years of selective pruning to develop an approximation; a tree that is on the path towards the "final" image. This is where you, the artist, must have faith, and patience, and visualize your developing creation as a real tree.

Sageretia, 2023

Proportions—tree height and trunk width

One guideline states that the bonsai tree height should be about six times the trunk width (diameter) measured near the base, to appear as a mature, old tree. However, **I don't concur with this**, because in nature we see countless trees of many species, almost none of which come close to such exaggerated proportions. Most trees are more than 20-times taller than their trunk width, depending on the species. **It's fine for a bonsai to have a thin trunk, as long as it looks like a tree in nature.**

Indoor or outdoor bonsai

You might have heard that there is no universal definition of "indoor" and "outdoor" bonsai. It entirely depends on where you live. For trees, the difference lies in each species' ability – or inability – to cope with outdoor conditions **in your climate**, compared to the natural environment of **the tree's native climate**.

To put it another way:

> *All tree species evolved outside over many millions of years, and during their evolution, there was no such thing as "indoors"!*

All trees evolved outdoors

So, in one simplified view, all bonsai are outdoor trees.

However, not all tree species evolved in your region, or in your climate.

Most species that we cultivate for bonsai evolved in another part of the world, with completely different climatic conditions and annual variations compared to where you live.

Autumn, mediterranean climate

A tree that is native to a tropical climate has not been exposed to cold winters; it is adapted to temperatures ranging between warm and hot.

The tropical growing season is all year round. For example, many Ficus species, such as the Ficus microcarpa (left), do not have the biological mechanisms to cope with freezing weather, and indeed will die if left outside in the frost or snow.

So, we often categorize these tropical species as "indoor bonsai". But this confuses things, because it depends entirely on where you live. An indoor tree in United Kingdom is likely to be an outdoor tree in India. Here in central Spain, I keep my tropical trees outdoors through summer, and indoors in winter.

For these reasons, it would be more accurate to categorize them as **"tropical bonsai"**. In autumn I bring the tropical bonsai trees indoors, as soon as the weather forecast predicts nighttime temperatures below 10°C (50°F). Our house is a little warmer than that at night.

If you live in the tropics, fortunately for you, you don't suffer cold winters. You can keep all your tropical bonsai outdoors all year round.

> **Note:** Outdoors in the sunlight, small trees in pots dry out much more quickly than large trees in the ground. A dry bonsai dies very quickly.

However, trees that evolved in colder climates do not survive for many years outdoors in a tropical location, because they evolved to go dormant every year through winter, in order to survive the cold period. Even if there's no cold period, they need their winter rest.

As you know, deciduous trees lose their leaves annually (right). They also convert their sugars to starches, storing the energy over winter; energy that is needed to push out new growth in spring. Coniferous trees also go dormant in winter, and their gummy resin acts as a kind of antifreeze against the cold.

In a temperate, cool climate, you must keep deciduous and coniferous trees outdoors all year round, so that they experience their annual rest period during winter; also to receive enough light for the rest of the year while growing.

> **Note:** During the growing season, spring to autumn, deciduous and coniferous trees can cope with being indoors for a few days.
>
> In winter, don't bring them indoors for more than a few hours. We don't want them to perceive warmer temperatures and "wake up" early.

If you keep deciduous or coniferous trees warm all year round, they survive for a couple of years, maybe longer, but gradually wither away and eventually die, because of their innate need to rest for a few months every year. Deciduous and coniferous trees are therefore decidedly "outdoor trees" if you live in a temperate or cold climate. But in the tropics, you need to refrigerate deciduous trees for a few months a year, in order to force the leaf-drop and the sugar-to-starch conversion.

Guidance by climate

The exact guidance **depends on the individual species**; however, as a general rule for most species you can follow these guidelines.

If you live in a temperate or a colder climate:

- Keep all your deciduous and coniferous trees outdoors.
- Keep tropical trees indoors in a bright window, unless the outdoor temperature is permanently above 15°C or 60°F.

If you live in a tropical climate:

- Keep your tropical trees outdoors.
- If you have deciduous trees or coniferous trees, store them somewhere in low temperatures for a few months each year, to create an artificial winter. Get them accustomed gradually at the start and end of the cold period each year, for example, by starting the change at nights only.

Summary:

If you live in a	Tropical climate	Warm, temperate or colder climate
Tropical bonsai (ficus, sageretia, etc.)	Outside all year	Indoors from autumn through spring
Broadleaf evergreens (olive, holm oak, etc.)	Outside all year	Outside all year; protect from cold in winter
Deciduous bonsai	Outside in summer; create artificial cold storage in winter	Outside all year; protect from severe cold in winter
Coniferous bonsai	Outside in summer; create artificial cold storage in winter	Outside all year; protect from severe cold in winter

> **Wherever you live,** if you plan to keep all your bonsai trees outdoors all year round, choose **species that are native to your region**, or at least from an origin with a similar climate.

Subtropical climates

If you live somewhere in between, like a subtropical zone that is hot for most of the year but still suffers cold winters, let the nighttime temperatures be your guide. If the temperature drops to about 10°C or 50°F, then be sure to bring all your tropical trees indoors, and keep all your deciduous and conifers outdoors.

Overwintering

If the temperature is freezing continuously for several days, consider sheltering your deciduous and coniferous bonsai trees somewhere that is "fridge temperature", to avoid the risk of ice starving the roots of liquid water, or expanding and breaking the pot. Your over-winter shelter should not get so warm that the trees wrongly perceive the start of spring and "wake up".

> **Note:** Observe the nighttime temperatures; this is the most reliable guide to your trees' winter requirements.

Notable exceptions

There are several tree and shrub species that can happily survive with all year-round warm weather **or** a yearly cold rest period in winter. Two notable examples are Chinese elms which are considered "semi deciduous", and olive trees which are broadleaf evergreens.

Two more exceptions: seedlings and cuttings in their first year. These can stay indoors for their first winter. Of course, deciduous and coniferous seedlings *can* stay outdoors in their native climates (remembering that as trees evolved, there was no such thing as indoors). But it could be prudent to give them a head start in life and protect them indoors in their first winter. Though, as mentioned, it is not strictly necessary, unless you live in an arctic climate.

The **Chinese elm** (right) is so resilient that it happily survives indoors all year without a dormant period, or outdoors all year, where in winter it can go dormant and lose some of its leaves. If you keep it indoors, be sure to place it in your brightest window: south-facing if you live in the northern hemisphere, or a north-facing window if you are in the southern hemisphere.

The **Olive** (left) also goes dormant in cold winter months, in as much as it stops growing, but it does not lose its leaves. It is very sensitive to low-light conditions, and therefore should not be kept indoors.

If the temperature stays below -4°C (25°F), below freezing for more than a day at a time, protect your olive bonsai in a greenhouse or a sheltered location that still gets daylight, but does not get warmer than fridge temperature. Bonsai trees in pots are more sensitive than their fully-grown siblings in the ground.

While deciduous trees can be kept in a cold, dark garage for the coldest couple of months in winter, this is not good for olives or indeed any evergreens which still need to receive minimal light through their foliage, even on the coldest days.

Tree talk for bonsai lovers

Let's take a brief look at the private life of trees. Trust me, this is not your average science class, but is "need to know" tree biology, like a bonsai owner's secret handshake. This knowledge, alongside the climate considerations in the previous chapter, helps us understand how our bonsai will respond to environmental conditions and to the work we do, like pruning and repotting. The essence of bonsai health.

Energy

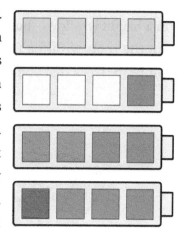

Think of a tree like a rechargeable battery. When the energy is high in the tree, it is much healthier and more robust. It grows vigorously and is more resilient to the harsh conditions and wounds imposed by its environment—or by the bonsai artist. Conversely, when the tree's energy is low, it grows weakly, if at all, and it is very susceptible to pathogens, insect infestations, fungal infections, drought, and physical wounds. A tree with low energy is more likely to die than a vigorous tree with a high state of energy.

The battery analogy is even more pertinent for a small tree in a container. The total amount of energy is much less than a fully grown tree; the growing space available to its roots is much smaller, and the total area of foliage – to absorb light energy – is much less.

With that in mind, don't do any major work on your bonsai when it is suffering, recovering from drought, or from an infestation and the after effects of an insecticide, or fungicide, or recovering from repotting. By "major work", I mean big branch chops, defoliation, or root pruning.

Light and leaves

A tree's leaves are like solar panels. When they receive light, they absorb the energy, and literally create sugars. This process is famously called photosynthesis. These sugars, dissolved in water, also known as photosynthates, are the sap that ultimately feeds all of the growing tree parts and provides the energy for cells to divide and create new growth.

In our battery analogy, light is the power source that recharges your tree, and enables it to grow. So, for all trees, we must ensure that they receive enough light throughout the year. And once again considering that all trees evolved outdoors, **if you keep a tree indoors you should position it in a bright window—but not directly above a radiator** which could dry out the roots within a fast hour. In winter, consider using LED grow-lamps for trees that you keep indoors.

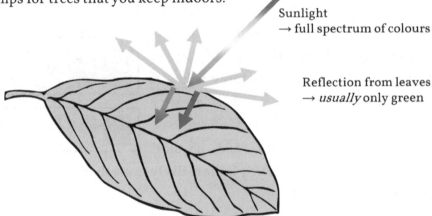

Leaves appear green because the chlorophyll in them reflects green light, and absorbs red and blue light. How do we know this? Because in autumn all the chlorophyll drains out of deciduous leaves, and we see the underlying colour of the leaves.

Keep deciduous trees outdoors in autumn because they need the shorter daylight hours and lower temperatures to prepare for dormancy. And, **always keep coniferous trees outside**, because their fine foliage needs direct unfiltered daylight all year round.

Photosynthesis and Transpiration

We all know that trees need light and water to grow, but they also need carbon dioxide, from air. That's where all the carbon comes from in trees; not from the soil, and not through the roots.

> *"stoma"* (singular)
> *"stomata"* (plural)
> → holes through which a leaf transpires

Let's look at a leaf at the microscopic level. Under leaves there are thousands of stomata—the holes through which leaves transpire.

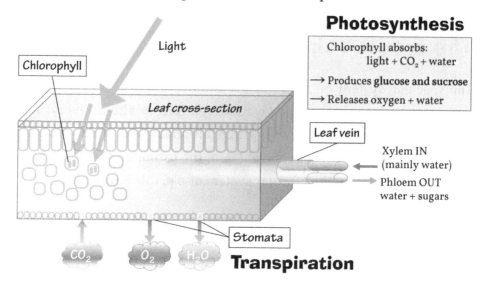

Here, you can see that leaf veins contain xylem and phloem tubes; the phloem is typically on the underside. When the leaf photosynthesizes, chloroplasts in the chlorophyll take in light, carbon dioxide from the air, and water from the xylem, and they generate the sugars which feed the tree's growth. Transpiration is the release of water vapour; leaves also release oxygen. This gas exchange is a bit like the leaf breathing through the stomata (though to be clear, it is very different from respiration).

Now, what happens to the leaf in the heat of the sun? Well, leaves are more than solar panels; they're also the cooling system for the tree. They suck water up the xylem in the stem, most of which evaporates to cool the leaf, and to cool the sap that goes back down the phloem. The evaporation increases in wind. When the roots start to get dry, the tree senses drought and closes the stomata to conserve water. Leaves 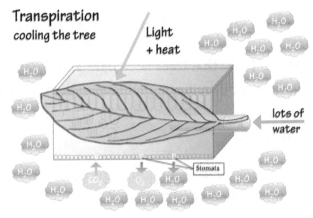 can dry very quickly and drop off. See *Reviving a dying bonsai*, page 154.

The tree's veins

The vascular system of a tree is profoundly important, and a basic understanding of these three layers is relevant to almost everything that we do with our bonsai. People often talk about "the cambium layer"; in reality there are three layers: phloem, the cambium and the xylem. These layers make up the vascular system of a plant. Simply put:

Phloem is a network of thousands of miniscule tubes that transport the sugars generated in the leaves down stems towards the roots. Phloem also transports sap from stored starch when needed, after it is converted back from starch to sugars.

Xylem is the network of thousands of tubes that take water and dissolved minerals from the roots up the tree. As the tree grows, older xylem lignifies, and becomes the inner heartwood, supporting the tree's weight and structure.

Cambium is a very thin layer of dividing cells between the xylem and the phloem. The cambium cells divide to create nearly all the tissue within tree stems, including the xylem and phloem tubes, and even the callus tissue that grows over wounds.

Roots, water and soil

Roots are the buried treasure of a tree. When they are healthy and vigorous, the tree can survive *almost* anything... within reason.

Roots grow by cells dividing then elongating at the root tips, and they absorb water and dissolved minerals from the surrounding soil. Over time, roots increase in thickness, stabilizing the tree in the ground. Thicker roots can store water and carbohydrates, but paradoxically, they have a lower capacity to absorb water compared to finer thread-like roots occupying the same volume. For this reason, in our bonsai, we try to **develop a dense mat of healthy, fine roots** beneath the soil line, rather than thick, woody roots. Thick surface roots may appear attractive and contribute to trunk base thickening, but under the surface we need fine roots to maximize water absorption efficiency in a small pot. See *Ten reasons to repot your bonsai*, page 129.

Water in the soil must be **fresh, oxygenated water to keep the roots healthy**. Water is oxygenated when in contact with air, and the oxygen helps fight any potential pathogens in the water and roots. This is easy to achieve if we use granular soil for our bonsai. Every time we water, gravity pulls the water through the soil and out the drainage holes, and that action sucks through some air too. See *Watering bonsai*, page 176.

By contrast, if soil loses its ability to drain, the trapped water without oxygen starts to stagnate, and grows harmful bacteria. This can rot the roots and could kill your tree.

Most granular soil retains less water than loamy mud or powdery peat moss. Therefore, for our bonsai, we **use soil that is granular, yet still retains enough water** to keep the roots damp between waterings. See *Soil*, page 161.

Granular soil

Plant hormones

Two natural hormones inside every tree, auxins and cytokinin, paradoxically both stimulate and restrict growth in different parts of the tree. Alongside other hormones, they work together, and yet as counterparts, to control and balance the tree's growth.

"Why's this important to a bonsai lover?"—you ask. Because it is these hormones that **switch on and switch off growth** in your bonsai as a **direct response to actions you perform**, like pruning the roots or branches, or wiring branches. Here's how they work:

Cytokinin is formed in the growing tips of roots, and it moves upwards encouraging growth in branch stems. And yet cytokinin is also a growth inhibitor within roots. Now imagine almost exactly the opposite:

Auxins are created in the growing tips of branches, and they encourage root growth. They move from cell to cell, down stems towards the roots. But guess what: while the branch tips grow, those same auxins inhibit the development of the lower buds under leaf petioles, called axial buds, and supress new lateral stem growth. Auxins repress back budding.

Apical dominance

Auxins make growing tips grow upwards while holding back lower growth. Because

> *"Apical dominance"*
> → trees grow upwards more than outwards

of this behaviour, **auxins are partly responsible for how tall trees grow**. They sense direct light and move away from it, concentrating on the shaded side of growing stems. And because auxins promote stem elongation, they encourage more growth on the shaded side of a stem than the sunny side. Knowing this, you can see why branch shoots naturally grow towards the light—even if you wire a branch horizontally, the tips grow faster on the underside (being shaded) than the top side, thus naturally growing them upwards. The top leader of a tree grows up towards the sunlight, because any slight sideways variation gets corrected and bent upwards by the presence of auxins.

There are also many types of woody shrubs that we use for bonsai, which have **weak apical dominance**. **Azaleas** and **Boxwoods** are two excellent examples. As shrubs, they grow laterally as much as they do vertically. You might say they have lateral dominance, although in reality it is more balanced. In these shrubs, the auxins are still produced at the growing tips; and cytokinin is still produced in the root tips; however, their distribution is much more even throughout the shrub than a tree. This means the inhibitory action of auxins on lateral buds is much weaker, so it is much more likely for those side buds to develop and grow sideways and outwards.

Growth cycle

Roots need to grow before stem growth can happen. Root growth creates cytokinin molecules, which move up the tree, where combined with auxins, stimulate stem growth. Stem growth produces more auxins, which move down to the roots, signalling more root growth.

When the cytokinin molecules build up in the roots, in higher concentrations than the auxins, the cytokinin signals sufficient or excessive root growth, and inhibits further root growth. Conversely, when the cytokinin concentration is lower than auxin content in the roots, the auxins signal more root growth.

Growth regulators

In this way, auxins and cytokinin have counteracting functions in any specific part of a tree, from branch tips to root tips. Yet they work together to balance the growth throughout the tree. For this reason, these hormones are often called growth regulators. Another growth regulator, **Gibberellin**, plays a big role in stem extension and flowering.

And for completeness, there are also other plant hormones. **Ethylene** ages and sheds leaves, flowers and fruit. **Abscisic acid** is a stress-response hormone, which responds to drought by closing leaves' stomata to conserve water, and in winter, induces dormancy.

Growth regulators and other hormones all work together like a signal network to regulate a plant's development, growth, and response to environmental conditions. Including any work that we inflict on it!

Rooting hormone

Commercially available rooting hormones can be used to encourage more vigorous root production. These are synthetically produced chemicals which simulate and augment the effect of auxins in the roots.

Whenever I propagate a tree via cuttings or with an air layer, I use rooting hormone, not only to increase the chances of rooting in the first place, but also to encourage a good root spread all the way around the trunk of the newly cloned plant. See *Creating an air layer*, page 126.

Painting rooting hormone on an air layer

Rooting hormone is available in liquid, gel, and powder forms. Gel is the most convenient. My most successful compound is the product of a few drops of the liquid in the powder, mixed with a few drops of water to make a thick balm that can be painted onto the target area.

There's no need to add rooting hormone to water for normal root growth of existing roots.

Note: Rooting hormone is very different from fertilizer supplements. See *fertilizer science*, page 85.

Part 2—Bonsai Reference A-Z

by concept or task

Air layer

See *Propagating trees*, page 123.

Apex

See *Creating an Apex*, page 80.

Automatic watering system

See *Watering systems compared*, page 180.

Bonsai benches/bonsai garden

It is simple to create a basic bonsai garden, that is not costly, looks beautiful, and supports all your potted trees during the growing season.

Small bonsai garden, small trees

If possible, position the bonsai garden so that your benches receive morning sun, and evening shade, especially in warm climates.

Another factor is the wind: if you live somewhere prone to strong winds, choose a position that is sheltered from the wind, to avoid broken pots and the heartache of wind-dried roots of a bonsai blown off the bench.

If you decide to position your bonsai garden in an area that is already paved or concreted, you can skip setting up the ground and go straight to the benches. Although, you might still decide to use gravel, as this can help the appearance of your bonsai garden.

If you are using an area of lawn or turf for your garden, you'll need to spend some time preparing the ground.

Bonsai ground area

For the bonsai ground, you need:

- Weed mats—large, thin polythene woven sheets that allow water to drain through, but prevent the gravel from bedding down into the underlying turf. A cutter.
- A long rope, or a garden hose, to initially define the perimeter.
- A long straight pole, and a spirit level.
- 15cm (6") ground staples, and a hammer.
- Gravel.
- A wheelbarrow and a rake.
- Sand, to help level the blocks.
- Optionally: a little cement to secure the lowest blocks.

Defining the perimeter

How to do it:

1. Use the stick and spirit level to ensure the ground surface is level both from north-to-south and east-to-west. Depending on the prior state of your ground, you might need to remove some turf or shovel it around to level it sufficiently.
2. Lay out the rope or hose to define the shape of your bonsai area.
3. Lay out the large weed mats. Hammer the staples down into the mats. Then cut the edges to the shape that you marked out.

Positioning the blocks | Weed mat

4. Measure out the positions of your base blocks for the two ends of your benches. Lay down some sand in those positions, and optionally some cement powder to prevent the blocks from sinking down into the sand.

5. Place the base blocks on the sand bases at the two ends, then ensure that the blocks are level with each other. If one end or one corner is slightly lower, lift it and add more sand under it.

   For the position and orientation of your blocks, see the *Bonsai bench* instructions below.

6. Lay out the gravel, and spread it evenly with a rake.

Bonsai bench

For each bench you need:

- 10 builders' blocks / breeze blocks.

Spreading the gravel

   Recommended block size: **40cm x 20cm x 20cm (16" x 8" x 8")**
- 6 wooden planks, preferably grooved to channel water away.
   Recommended plank size: **180cm x 20cm x 4cm (72" x 8" x 2.5")**
- 8 long wood screws + 8 raw plugs, a drill, and a screwdriver.
- A wood saw, only needed if the length of planks is greater than your desired bench length.

How to do it:

If you have already laid out the base blocks, skip straight to step 2.

1. Measure out the positions of your base blocks for the benches. Place the base blocks and ensure that they are exactly level with each other. If you are placing them on concrete or paving, still check that they're level because the bench should be horizontal.

   Assuming your blocks are exactly twice as long as they are high, I recommend one block on its side and one block vertically. Then, add another block on top of the one on its side. This makes three ground blocks at each end of your bench.

2. Place four of the planks on the blocks, exactly in the positions ready for the lower shelf of your bench.

3. Drill tiny holes through the wood and into the blocks at each end of each plank. If necessary, insert raw plugs into the holes in the blocks, then screw the planks onto the blocks.

4. Position two more blocks, one on top of the other, at each end of the rear of the first shelf. This uses the last four of your blocks.

5. Position the remaining two planks as your top shelf.

6. Drill and screw the top planks into place.

Bonsai garden

Optional extras

Extra shelf: You can add a near-ground level shelf in front of, and lower than, your previously lowest shelf. Lay two more blocks on their sides at each end in front of the bench, then place one plank across them.

Shade support: If you live in a hot, dry climate, for the summer months you might need to install a large 50% shade cloth over the benches, to reduce the power of the direct overhead sunlight.

Extra shelf, and 50% shade cloth above

You can add wooden beams vertically to the end of your bench, screwing each one tightly to the building blocks in two or three places to secure it firmly in place. This gives you the basis for tying the shade cloth above the bench. You might need to add guy ropes to each corner of the shade cloth, in order to keep it taut and prevent it sagging onto your trees.

Automatic watering system: If you plan to install an automatic water system, the rear side of the wooden planks are an excellent location for fixing some brackets that hold the wider tubes of your watering system. Try to choose brackets that are securely fixed with at least one screw, and yet easily allow insertion and removal of your water pipe.

Then, on the benches themselves, you can also screw on some metal supports to hold the thinner tubes which take the water to the nozzles. See *Automatic watering system*, page 183.

Bonsai sizes

Aside from innate beauty, part of the appeal of bonsai is their small size.

Classical size names

Although there are several classic names for different sizes of bonsai trees, there are also **many inconsistent definitions** and **conflicting classifications** out there. Although the names are derived from Japanese, interestingly most Japanese bonsai artists are not too concerned about the exact size name, and focus more on the "feeling of the tree". For example, if you can hold it in one hand, you have a feeling for its approximate size and weight; the exact height is less important.

Portability scale: practical size names

In practical bonsai work it is more useful to consider the sizes in terms of how easy each is to carry, to keep watered, to prune, or to repot. For this reason, I have added a portability scale, which gives an idea of how portable and easy-to-maintain each size is.

Classical sizes with a portability scale

Other sizes also exist outside these height ranges; however, in bonsai literature, forums and classes, these sizes are most commonly used:

Classical size name	Height range	Portability scale
Mame	5 – 15cm (2 – 6")	Palm-top tree
Shohin	12 – 20cm (5 – 8")	One-hander tree
Komono	15 – 25cm (6 – 10")	One-arm tree
Katade-mochi	25 – 45cm (10 – 18")	Two-arm tree
Chumono (chiu, chu-hin)	40 – 90cm (16 – 36")	Two-person tree

These size ranges overlap somewhat, and it can be difficult to exactly categorize a tree, for example, between a shohin and a komono. It is **more important to consider the practical implications** of different sized trees. For this reason, I use the classic names very loosely as nicknames, because for example it's quicker to say "shohin" than "one-hander tree".

Care requirements vary considerably for different bonsai sizes.

Special attention for different sizes

Here is a size-related care guide to help keep all your bonsai trees alive despite a busy schedule, yearly repots, hot summers and travel away.

Palm-top bonsai: Mame

Mame is pronounced *ma-me;* not *maim* as my kids enjoy saying. These miniature beauties fit easily in the palm of your hand.

Over the years, I have sadly lost several mame-sized trees. Because of the tiny pot, a vigorous mame can dry out within an hour on a hot, sunny day. It's critical to keep the soil damp, otherwise your tree could die the same day. The tiny trunk and roots don't store much water.

Mame Sageretia theezans

 Keep a close eye on your mame trees in tiny pots, and **never let the soil dry out**. Keep them in **semi-shade** on sunny days.

Watering: Use a small fine-spout watering can. On hot days be prepared to water trees several times. To slow the drying rate, place all your mame pots on a humid platform such as a drip tray, or a thick bed of wet moss.

 Place each pot on a humidity tray with gravel, to retain a pool of water under the pot, without "drowning" the roots.

When I travel away, I leave the tiniest trees on **water boxes** for the duration, with a wick dipping into the water. See *Water box*, page 182. Water boxes work well for a couple of weeks. Don't leave them in sunlight because the water gets hot and could cook the roots. If you use a water box all summer, refresh the water in the box and soil every week.

Pruning: for the first few years in a tiny pot, mame bonsai need to be pruned frequently, to keep growth compact and to limit root growth.

Special care and position: A mame will die rapidly if a curious pet claws the roots out of the pot. Also, tiny pots can easily get knocked off a shelf.

If you have curious or playful pets, protect your trees by **wiring drainage mesh over the soil surface**. Place the pots in a location where they won't fall and break if they get knocked.

Repotting: Mame trees need to be repotted yearly, because their roots can fill out a tiny pot in less than a year. The good news is, with their tiny root system they are quick and simple to repot, and because of their minimal total leaf mass, you can prune the roots very substantially.

Mame *Portulacaria afra*

Use highly retentive soil granules of 2.5mm (1/10")—like pine bark (sieved) and akadama. It's okay if the soil doesn't uphold its granular structure because you repot it again within a year.

Note: *Portulacaria afra* makes an excellent species for a mame bonsai, because, being a succulent, it withstands dry conditions much better than a tree. You won't need to water it multiple times a day.

One-hander bonsai: Shohin

This size might be the most commonly available category of bonsai.

Watering: While not as small as mame, shohin-sized bonsai still need frequent watering because of their small pot size. Just like mames, when you travel away, you can place indoor shohin-sized bonsai on water boxes, in the shade.

Pruning: Prune the canopy in late spring to reduce the foliage area before summer.

Shohin Chinese privet

Repotting: A relatively quick job for a bonsai this size. To do a thorough repotting, set aside an hour, or two if it's your first time. The frequency of repotting depends on the species, but if you frequently prune the canopy, you can leave most trees for two years before repotting.

Slow rooters like boxwoods and dwarf pomegranates can be left three years between repots. Species known for super-vigorous rooting, like crab apples or trident maples, need yearly repotting. You can extend the time by slip potting or using an oversized pot, but be aware that the root-pruning will be substantially more complicated after two years.

One-arm bonsai: Komono

I call this size "one-arm" because you need a strong hand and your forearm to hold the tree, or brace it against your stomach. Care guidance for maintenance and repotting is the same as with two-hander trees.

Komono-size (and larger) trees with a lot of foliage evaporate off pints of water on a hot day, and even more in direct sunlight.

Pruning: Remember that more foliage surface area needs more water.

 Be sure to prune the branches in late spring or early summer.

If the branches don't need pruning, and the tree is healthy and vigorous, consider instead defoliating in early summer. See *Defoliation*, page 81.

Two-hander bonsai: Katade-mochi

When you carry a tree this size, it might cover your entire face as you walk around; don't trip and break the pot—or worse, a bone!

Repotting: It can be a lengthy chore to do a thorough job of root pruning. Set aside a few hours, including preparing beforehand, and clearing up afterwards. The frequency of repotting depends on the species, but in general if you prune the canopy quite frequently you can leave most trees for 2-3 years before repotting. If the surface soil particles have broken down, or have become overrun with moss or weeds, considering doing a "pot refresh" which takes a lot less time. See *Pot refresh*, page 122.

Slow rooters can be left for more than 3 years without pruning the roots.

The most vigorous rooters need root pruning every 1-2 years. If you're not sure, in Spring, ease the root ball out of the pot and see if it is rootbound. See *Root pruning*, page 142.

Watering: You can get away with watering large trees in big pots less frequently than small ones, because the trunk, roots and pot hold more water. Although, when you water these big bonsai, they evidently need more water than a small bonsai. As always, keep the water flowing until it drains through the holes on the base of the pot, partly to ensure that all the soil is wet, and also to check that the pot is draining freely.

Bonsai soil

See *Soil*, page 161.

Bonsai styles and styling

While this book primarily focuses on the horticultural aspects rather than advanced design matters of bonsai, this section offers a cursory glimpse into my tips on designing and developing various bonsai styles.

Bonsai design

Here are a few foundational design principles that can serve as rules of thumb for all bonsai artists:

- Choose the **tree's front**. See *Front of tree & planting angle*, page 108.
- Your bonsai is a 3-dimensional tree; **consider its depth**, with front-and-rear movement and branches, as well as left-to-right and height.
- **Use triangles as your approximate pruning guide**. As a general rule, prune the canopy to an irregular triangle shape. Also, considering the 3 dimensions, triangles become cone-shaped. The irregular-triangle guide is also used to shape multi-trunk bonsai and forests.
- **Branches should be thicker near the trunk**, and progressively thinner as they divide out to smaller branches and twigs.
- **Tree sections should be longer near the base**, and progressively shorter moving higher into the canopy or further from the trunk.
- **Asymmetry and balance**: a perfectly symmetrical tree can appear unnatural and uninteresting. Branches should not mirror each other left to right. Leave spaces in the canopy to counterbalance branches.
- **Design harmony and unity**: all the parts of a bonsai should follow the same style, to invoke feelings of harmony, comfort and repose.

For more detailed design principles, and style variations and nuances, I recommend the book *Principles of Bonsai Design* by David de Groot.

Bonsai styles

Let's look at some of the formally recognized styles, with my tips on cultivating and maintaining them. There are many variations and other styles, and in practice, most bonsai are a blend of more than one style.

Broom style *"Hokidachi"*

Broom-style Zelkova, refined

The broom appears deceptively easy to develop, because in a simple scenario you can just let a tree branch out in all directions, and profile trim around the shape of the canopy each time you prune it. Just like hedge pruning, it keeps the tree in shape and encourages more ramification. See *Structural pruning versus profile trimming*, page 79.

However, a well-developed and refined broom-style bonsai has its primary branches well positioned and proportioned relative to the trunk in order to be a truly convincing portrayal of a tree.

Broom-style Zelkova in development

So, while you develop the bonsai, it needs structural pruning on the primary branches. On a large tree, this can go on for ten years or more; on a small shohin-size bonsai, plan about five years before you can revert to profile trimming alone. Also, as most trees are apically dominant, the vertical and higher branches tend to thicken faster than the lower, horizontal branches.

 Prune back vertical and higher branches more aggressively than the lower, horizontal branches.

There are several variations of broom style: flame style, globe style, umbrella style, and the "true" broom style like an old garden broom.

 Decide at the outset which style you'd like to develop. It can be difficult to change part way through development.

Curiosities of the broom style

The broom is unique amongst bonsai styles, in that we often see inverse taper in the trunk—it can widen leading up to the first spread of branches. This accurately reflects the growth of many trees in nature. However, the branches should taper to thin twigs, like any other bonsai.

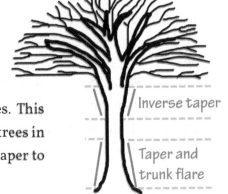

The broom style is suited to deciduous trees, because their fine ramification can look spectacular in winter when they are bare of leaves.

The canopy of a broom covers the view of the trunk from all directions. Other styles leave a window in the canopy to expose the beautiful trunk.

Informal upright style
"Moyogi"

This is the most commonly seen bonsai style, and probably in nature too. The trunk has some visually interesting movement, and is conically tapered.

Branches should emerge from the outside of trunk bends—though it's not always possible.

Informal upright Wild olive

Each section of an informal trunk should be both shorter and thinner than the section beneath. Despite the trunk movement, the apex should be directly above the trunk base, and for aesthetic appeal, the upper section should slant in the same direction as the first trunk section.

Informal upright Sageretia — diminishing trunk sections

Tips and tricks:

Apically dominant trees need their upper branches pruning much more aggressively than the lower branches, thus contributing to trunk taper over time. See *Ten reasons to prune your bonsai, reason 10* on page 68.

To develop an informal upright trunk, you can achieve a beautiful zig-zag movement with a **series of trunk chops over many years** (10-20 years for a large tree; only 3-5 years for a tiny mame-size bonsai). For each chop, leave a small shoot in the right place to grow as the next leader. See *Trunk chops, trunk thickness and trunk taper*, page 172.

Two disadvantages of trunk chops to develop a bonsai are the sheer amount of time it takes, and the challenge of healing the chop wounds.

Alternatively, to develop informal movement in a shorter time, **wire the trunk of a young whip**—a seedling or cutting in its second year. Beware of some major disadvantages though—see *Wiring the trunk*, page 196.

Curiosities of the informal upright style

Informal upright is the style that is most combined with other styles. For example, a "root-over-rock style" bonsai refers to the planting, but the tree itself is most commonly informal upright style. Similarly, multi-trunk styles are often a variation on informal upright.

Formal upright style "*Chokkan*"

Formal upright Cedar

We usually use coniferous species for this style, to mimic tall conifers in evergreen forests. Cypress, juniper, larch, pine, and spruce are all suitable.

The formal upright is easy to visualize but classically difficult to develop well. The trunk must be flawlessly straight with no movement, and it must taper somewhat from base to tip. Branches typically grow horizontally all around the trunk, or even sloping slightly downwards, to appear like vast conifers with heavy branches. To achieve this, we usually need to wire the branches downwards moderately. Find a picture of a tall conifer in the wild, and mimic the angles and lengths of the branches relative to the trunk.

Slanting style "*Shakan*"

Slanting-style Maple

This style can have either a straight or curved trunk. Slanting at an angle between about 15° and 45° off vertical, the apex is seen to one side of the trunk base from the front view. Any further than 45° and the trunk is closer to horizontal than vertical (which would practically be a semi-cascade).

Slanting-style bonsai trees sometimes have an asymmetric first branch which gives a visual sense of balance and repose.

Tips and tricks:

- **Wire branches horizontally** on both sides of the trunk, so that the tree appears to have always grown at a slant, rather than the "fallen tree" look, characterised by branches sloping downwards on one side of the trunk and sloping upwards on the other side.
- **When repotting**, wire the trunk base firmly in to the pot, for 4 to 6 months, until the roots have extended firmly into the new soil.

Twin-trunk style *"Sokan"*, triple-trunk *"Sankan"*, multi-trunk and clump styles *"Kabudachi"*

In a bonsai with more than one trunk, to appear natural, all of the trunks should have the same basic style. Each trunk should be a distinct size.

Twin trunks should have a dominant (primary) trunk and a secondary trunk that is smaller and slightly leaning away from the primary trunk.

Triple-trunk elm

Similarly, triple trunks should have a primary trunk, and two smaller trunks with distinct sizes, leaning away at different angles.

Twin-trunk and multi-trunk trees occur commonly in nature with certain species such as birches, or holm oaks. Trunks emerge from the trunk base – or below ground – and not above a length of single trunk; otherwise they are thick low branches rather than trunks.

Tips and tricks:

- When **developing a bonsai with multiple trunks**, as long as the tree is young and supple enough, consider wiring the trunks so that their movement is in harmony with each other, and to lean the second and third trunks slightly away from the primary trunk.

- If the tree doesn't have roots emerging from the join of the trunks, **bury its lower trunk** beneath the soil surface for a few years. This encourages roots from the trunk join, so that it becomes one large trunk base, instead of thick branches dividing from a single trunk.
- If it is a **strongly developed branch division** above a length of trunk, consider applying a ground layer just beneath the split, to force roots to emerge there. A year later, you can remove the lower trunk. A ground layer is an air layer near ground level, which is often successful, but risks the tree. See *Propagating by air layering*, page 126.
- When pruning a bonsai with multiple trunks, treat it as one canopy. The lopsided-triangle shaping rule applies to the overall tree.

Forest style "*Yose-ue*"

A bonsai forest is a group planting of multiple trees, which are usually the same species and style. They can be formal- or informal-upright style. Like multiple-trunk bonsai, forests should be comprised of a primary, dominant tree, plus a number of smaller trees. The primary tree should be off-centre, so that the forest is asymmetrical.

Juniper forest or spinney

The spinney, or grove style, is similar to a forest, except usually with fewer trees, placed more closely together.

Tips and tricks:
- **To create a forest**, choose a wide, shallow pot. Study your trees, and before potting, first **create your planting design on paper**.

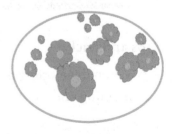

- **Draw taller, thicker trees nearer the front**, and successively smaller moving back and to the sides. This forms an illusion of perspective, giving the impression of viewing the forest from nearby.

- To exaggerate the effect of nearby perspective, consider potting the outer-most trees at angles leaning outwards; trees to the left of the planting lean to the left, and those at the right side lean right.

Ginkgo Biloba forest with nearby perspective

- Alternatively, you can design the opposite effect, with similar sized trees throughout the forest, all at the same planting angle, to appear viewed from a greater distance.

- To **prepare the pot for a forest**, before adding soil, attach many wires to the bottom of the pot through the drainage and wire holes.

  If there are no wire holes, create a grid of thin wooden sticks tied down through the drainage holes. Tie wires to the grid, ready for the trees.

  Pot with grid and wires

- Add bonsai soil and mound it towards the middle, with the highest point off-centre. If your forest has many trees, such as fifteen or more, consider two mounds with an off-centre dip in between.

- When **first potting a forest**, start with the largest tree, off-centre, then move outwards planting successively smaller trees.

- **Repotting a forest** should not be more frequent than every three years. As long as you're happy with the forest layout, repot it as one large root mass, trimming the long roots and around the perimeter.

- When **pruning a forest**, start by profile trimming the overall canopy to the lopsided triangle shape. Then focus on individual trees, ensuring branches don't grow into each other. Prune off low branches to reveal the lower trunks.

Cascade style *"Kengai"*

Pine cascade

About half the visual mass of the tree should be below the level of the pot rim. Junipers and pines are the most used species for cascade style while their branches are flexible and can withstand quite extreme wiring.

The lowest branches grow with naturally less vigour than the upper canopy, since most trees are apically dominant. Due to more auxins concentrating in shaded areas of stems, branches naturally grow upwards towards the light. For this reason, it's important to wire the lower branches with the foliage pads fanning outwards so that the thin foliage receives as much light as possible. See *Apical dominance*, page 36.

Tips and tricks:

- To **develop a cascade**, wire the trunk of a young whip and bend it down with curves so that branches emerge on the outside of bends. See *Wiring the trunk*, page 196.
- Wire one branch as an upward leader so that the tree develops a crown above the cascading canopy.
- As with all wiring, check frequently for signs of the trunk or branch thickening. Remove and rewire as soon as needed.
- Wire the pot tightly to your bonsai bench. Cascading trees in tall pots blow over in the wind more easily than most bonsai.
- **Prune the upper crown** harder than the cascading branches.

Young juniper wired to cascade

Semi-cascade style *"Han Kengai"*

Semi-cascade style trees reflect the natural growth of trees on steep mountainsides or cliffs. The trunk angle is below 45°, and the lowest branch tips are more-or-less level with the trunk base. Most species can be grown as a semi-cascade, except for trees with strongly upright growth.

Semi-cascade Sageretia

Windswept-style boxwood

Windswept style *"Fukinagashi"*

There are formally two styles: windswept and windblown.

Windswept style is the depiction of a tree that has been shaped by a constant prevailing wind over decades or centuries. Only branches pointing in the downwind direction survive in the long term, and any shoots in other directions die or get bent by the wind.

Windblown style is like a snapshot image of a "normal" tree in a strong wind. Branches emerge from the trunk in all directions but as they divide and subdivide, the secondary and tertiary branches bend towards one direction, as if they are briefly blown that way by a strong gust. In practice, windswept trees also appear windblown.

Tips and tricks:
- Use spiral wiring. Don't use guy wires, which pull the tree down on one side, gradually tilting the trunk and skewing the roots.

- After repotting, place a heavy rock on the soil the "upwind" side of the trunk (left-side in the previous photo), to hold down the roots for a few months, balancing the tree until those roots fill out the pot.
- While not on public display, keep the whole pot tilted up to an angle of about 20° so that the foliage pads point upwards slightly while they grow. That way, when the pot is set level again for display or for photography, all the new growth appears horizontally windblown.

Landscape *"Penjing"*, *"Saikei"*

Saikei style is a group planting in a landscape that usually includes rock plantings with multiple trees and other elements like sandy pathways, and moss to mimic shrubbery and wild grass.

Penjing is a Chinese term for a tree planting; it differs from the Japanese "bonsai" in that a penjing includes additional elements that make it appear as kind of live scene. For example, a tree planting with a statuette, like a scale model animal and farmer, or fisherman. A penjing can also depict water-side scenes, with rocky beaches, or cliffs.

Tree styles in landscapes are usually informal upright, occasionally cascade clinging to a cliff, or sometimes windswept.

Other formally recognized styles

- **Abstract style**—styled with artistic license, with somewhat unnatural tree forms like "floating clouds" or "dancing dragon".
- **Banyan style**—with many air roots hanging from branches.
- **Driftwood style**—the deadwood is the focal point of the bonsai.
- **Free style**—a bonsai styled without adhering to a formal design. It can appear like a blend of various different styles.
- **Hollow-trunk** and **Split-trunk style**—mimicking old trees that have lost a significant part of the trunk's heartwood to decay.
- **Literati style**—a tall, slender trunk with some subtle movement, relatively few branches and sparse foliage. The bark should appear old and mature.

- **Raft style**—mimicking a fallen tree, growing several upright trunks along its length. Use a line of branches to grow the new trunks.
- **Root-over-rock style**—choose a hard, textured rock that won't crack in frost; use raffia or wire to tie the roots while still supple. Over a few years, successively reveal the roots from the top down. Exposed-roots style is similar, but with the roots bound more loosely so that the supporting stone can be finally removed.
- **Rock planting**— Also known as tree in a rock, or clinging-to-a-rock style. The key to a rock planting is the quality of the rock, like a cliff or a rocky outcrop, with indentations that can be used to plant trees.
- **Twisted-trunk style** and **Coiled-trunk style**—the twisted trunk often shows alternate dead wood and live wood, twisting up the trunk. The coiled trunk is an exaggerated version, showing more deadwood, and hollows where deadwood has completely rotted.
- **Weeping style**—to mimick a weeping willow or similar tree, wire many long shoots and bend them downwards like a fountain.

New styles

Some styles are not formal or traditional Japanese styles, but are often discussed in bonsai talks, forums and social media.

- **Fairytale style**—movie-like forms, sometimes spiky or gnarly and contorted, carved with ghostly shapes and hollowed trunks, or stumps with personality so hideous that they're beautiful.
- **Naturalistic style**—trees that are styled to imitate the natural growth habits of their species, without following a formal style.
- **Orchard style**—styled as a vase-shaped fruit tree.
- **Pierneef style**—styled as an acacia tree in the African savannah.
- **Yamadori style**—twisted and contorted, appearing as battered by severe conditions in high mountains. Lots of dead wood.

> **Note**: The Japanese term "**yamadori**" traditionally refers to material that is mountain collected then styled as a formal style. However, the more recently termed "yamadori style" can start from a young whip, wired and coiled tightly to form the exaggerated twists.

Bonsai tools

Have you ever wondered why we use bonsai-specific tools instead of cheaper alternatives from your local DIY hardware store? And why do bonsai tools have a unique appearance, with large handles and tiny blades? It's not just for aesthetics; it's all about finesse and leverage.

Leverage

Consider a branch cutter with a handle length of 25cm (10"), and the cutting blade at only 2.5cm (1") from the pivot. This tool magnifies your hand's force by a factor of ten. So, if your hand squeezes with a force of 20kg (44lb), the blades exert a cutting force of 200kg (440lb). That considerable force, coupled with a meticulously sharp blade, allows you to cut through extremely hard wood.

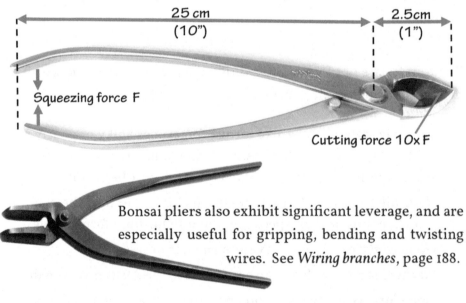

Bonsai pliers also exhibit significant leverage, and are especially useful for gripping, bending and twisting wires. See *Wiring branches*, page 188.

Right tool for each job

Keep in mind that old deadwood can be much harder to cut than live wood. To cut through thick deadwood, use a narrow branch saw. This is very useful for thicker branches and trunk chops—see *Chopping the trunk*, page 173. A branch saw makes a straight cut, leaving a flat plane of wood, and also avoids blunting your fine bonsai cutter.

Stainless steel or blackened steel?

Both stainless steel and blackened steel are resistant to corrosion, to some extent. When choosing, consider these differences:

Stainless steel tools are durable and very corrosion-resistant. They can cost more than the similar blackened steel tools, because of the alloying elements and manufacturing processes. However, stainless steel is arguably worth the extra money, because its durability and corrosion resistance are superior to those of blackened steel. On the surface of stainless steel, a naturally formed layer of chromium oxide protects the metal from rust, and self-repairs minor surface scratches. This means that fine, sharp blades are less prone to scratches and rust.

Blackened steel tools are sometimes preferred by artists for their initial sharpness. Heat-treated high-carbon steel is theoretically harder than stainless steel and can therefore achieve a finer cutting edge.

However, this doesn't necessarily help us, because we don't know the grade of steel or its carbon content in most commercially sold bonsai tools. Blackened-steel tools are usually surface treated rather than high-grade steel, so the potential advantage is irrelevant.

In practice, stainless-steel cutting edges tend to maintain sharpness for longer, at a comparable price.

I use stainless-steel cutting tools, and blackened-steel pliers. In my tool box, a blackened-steel root scissor has rusted; this has never happened to my stainless-steel tools. I also have to oil the pivots on my blackened-steel pliers so that they're not stiff. I don't need to do this with my stainless-steel tools.

Sharpen your tools

Make it a yearly practice to sharpen each cutting tool. Use a wet stone, available from bonsai shops and some hardware stores.

The blade's sharpness does not have to be perfect, but you will notice the difference when cutting. A sharper blade makes cleaner cuts which are less prone to pathogens entering the organism than jagged, dirty cuts.

Basic starter tool set

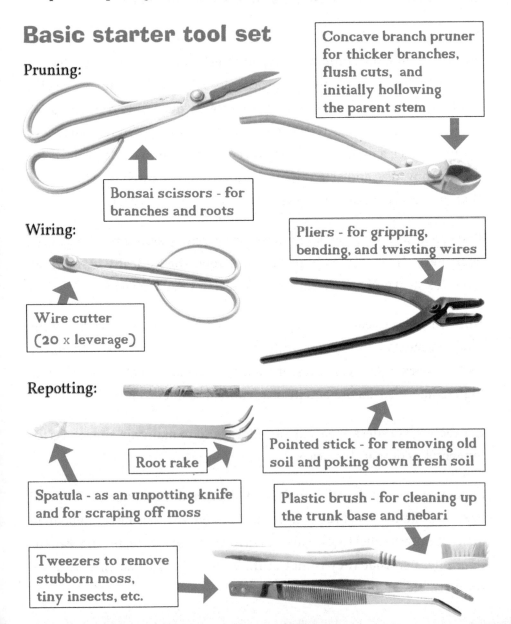

Pruning:

Bonsai scissors - for branches and roots

Concave branch pruner for thicker branches, flush cuts, and initially hollowing the parent stem

Wiring:

Wire cutter (20 x leverage)

Pliers - for gripping, bending, and twisting wires

Repotting:

Spatula - as an unpotting knife and for scraping off moss

Root rake

Pointed stick - for removing old soil and poking down fresh soil

Plastic brush - for cleaning up the trunk base and nebari

Tweezers to remove stubborn moss, tiny insects, etc.

Branch pruning: why, when, how

When setting out to prune a bonsai, we need to understand the reasons for pruning, and what we're trying to achieve by pruning our bonsai.

Ten reasons to prune your bonsai

1. My goal with bonsai is to make it look like a believable tree out there in nature; **a miniature representation of a real tree**. And that is our first reason for pruning bonsai, because we want to keep the form or the shape of the bonsai looking like a tree in nature, and not just an unwieldy bush or a stick in a pot.

2. Related to the first reason, we want to **achieve a smaller leaf size**, so that the leaves appear in realistic proportion to the overall tree size. And so it adds to the more believable image of a natural tree.

   If you look at a long stem you'll nearly always see that the first few leaves down near the base are smaller leaves, and as the shoot grows, the leaves get progressively bigger. Similarly, the internodes are shorter down there, and grow progressively longer.

Shorter internodes at the stem base

We can achieve this goal of reducing the apparent leaf size of the tree, by cutting each shoot nearer to its parent stem or the trunk.

3. By trimming a shoot at any point in its growth, we **encourage finer ramification and more compact growth**. So, we prune to develop progressively finer branching around the cut points.

   And over a long period of time, when your tree has a lot more ramification, meaning more fine branching and "twigginess", once again you're positively enhancing the overall image of a real tree.

4. We prune branches to **create growing space** within the canopy of the tree. This allows more light and air between areas of compact growth. Also by doing this, we create distinct **foliage pads,** which are attractive and mimic clustered masses of foliage on full-size trees.

   Equally important, trees need to grow in order to stay healthy, so when we prune the silhouette to shape, we also need to allow space for new shoots to grow within the canopy. Interior, shaded branches with no new growth will gradually wither and die.

5. We prune to **encourage more back budding** on the tree.

   Back budding happens when a plant creates new buds on older wood, nearer the trunk or trunk base. These are also called adventitious buds.

   Explanation: As a branch grows longer, the cells divide in the growing tip, releasing auxins, which move down the stem, acting as a chemical signal to inhibit lateral growth. Think of it like a red

   Back budding on the trunk

   stop light, halting the growth of new buds, and inhibiting buds from developing into secondary branches.

By pruning a branch some way along its length, we cut off the backward flow of auxins in that stem, meaning we remove the "red stop light" signal that inhibits lateral growth. With fewer auxins present in the stem, latent buds can get activated by a higher concentration of cytokinin, to push out new lateral shoots. Auxins and cytokinin are explained in detail in *Plant hormones* on page 36.

6. In order to avoid ugly bulges growing at branching points, we need to **remove new shoots growing from the crotches of branches**. If left to grow, these gradually form a knuckle-like lump which, on a bonsai, looks a bit ugly, but more importantly the lump becomes way out of proportion on the parent branch, compared to the joints on a full-size tree.

7. To **remove awkward branches** or visually jarring branches, for example, disproportionately huge compared to the rest of the tree; vertical shoots growing directly upwards or downwards; or branches that cross over each other—prune back one of them.

8. To ensure that our **branches are always bifurcating**, that is, there are only two stems emerging from each branching junction.

    On the trunk, this means each primary branch should emerge from a non-shared point on the trunk, and ideally at a unique height.

    On a branch, this means one branch divides to two; those two divide to four; those to eight, and so on.

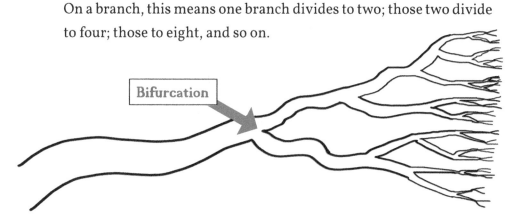

Notice also how the lengths of the sub-branches diminish as the branching gets finer. This is another realistic feature of how trees grow in natuure.

Moreover, we want to avoid the development of inverse taper on any of the stems. Just like the trunk, branches look more natural when they taper from thicker to finer. When we have multiple shoots growing from the same node, the stem sends more sap to that node than to others, and over time the branch thickens just there more than it does lower down. By ensuring the branches bifurcate, we enable a balanced flow of energy throughout the entire branch and its sub-branches.

So, choose two shoots to keep growing, and remove the other shoots emerging from the same node.

9. To **keep all the parts of the tree in proportion** with each other, with respect to the relative proportions in a tree in nature.

10. To **keep the branches smaller near the apex**, while keeping lower branches larger. This lets more light to penetrate through to the lower branches.

If we allow the apex branches to grow enough to completely overshadow the lower branches, then these gradually decline in vigour while the apex continues to thrive. Growth in the lower branches would become sparse and loose, and back budding less likely.

Prune the apex more than the lower branches

In general, most trees – broadleaf and conifers – have apical dominance, while most shrubs like azalea and box are laterally dominant. See *Apical dominance*, page 36. It's more important to hard prune the top growth of apically-dominant trees. Prune the apex of laterally-dominant shrubs and trees somewhat less, since top growth tends to be weaker.

Oh, and did I say there are ten reasons to prune a bonsai? There are more! Here's an eleventh reason:

11. We prune our bonsai to try to **achieve more trunk taper**, so that the trunk is wider at the lower part and progressively becomes narrower up the trunk.

Trunk taper

And over a long period of time, we can help achieve that by retaining more foliage in the lower branches which gives more energy to the lower trunk, and continually trimming higher up so that there is less vigour there. Effectively, we are directing more energy to the lower part of the trunk than to the apex, so that the lower trunk thickens at a faster rate than the upper trunk. See *Trunk taper*, page 174.

Exception: in the umbrella variation of the broom style, we usually see inverse taper on the trunk, because many branches naturally emerge from the same height on the trunk. See *Bonsai styles*, page 51.

When to prune

Firstly, you can prune **tropical trees** at **any time of year**—whether you keep the tree indoors, or you live in a tropical region where there are no cold winters. See *Indoor or outdoor bonsai*, page 25.

However, trees that are native to temperate or colder climates – deciduous trees, conifers, and many broadleaf evergreens – all have an annual energy cycle that evolved to make the most of the growing period yet also to withstand freezing winter temperatures. We need to respect this energy cycle in our bonsai, both for pruning branches discussed here, and for pruning roots—see *When to repot*, page 136.

The energy calendar

You've probably heard of summer pruning and winter pruning. Let's study the seasons and see when the best time to prune is, and why.

Be aware that this calendar is very generalized by season, and all species and cultivars have slightly different needs and timetables. It would be a mistake to prescribe a month-by-month schedule because the right time for an activity in your climate is likely at a different time in my climate—even between microclimates within the same country.

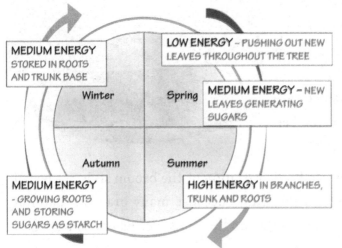

Consider the energy level of the tree. When the energy is high, the tree has reserves; it has plenty of sugars stored and moving inside it.

So, you can go ahead and prune, and it's going to push out new branches or new leaves after pruning. Thus, when the energy is high in the tree it's a good time to prune. When the energy is low in the tree, pruning off large branches removes significant energy. Furthermore, pruning cuts can weaken the tree, reducing chemical defences and allowing pathogens into the live cambium and phloem. Therefore, prune at high-energy times so that there are enough reserves to activate new growth, to divide cells and to create all the defence chemicals and callus tissue.

Summer

In summer, the energy is high because the leaves are generating copious amounts of sugars; enough to grow new shoots and roots, and at the same time withstand the loss of a large part of the tree. So, **summer is a good time to prune branches.**

Autumn

During autumn, the energy is somewhat lower, because there is less light, and the leaves are turning yellow on deciduous trees. Evergreens also slow down their sugar generation ready for winter dormancy. Trees still photosynthesize in autumn, but much less so than summer.

Zelkova in autumn

Pruning in autumn can be detrimental in two ways: firstly, it can hinder the generation of new buds. And secondly, just before leaf fall, the tree is trying to store away as much as possible of the photosynthates, packing them down the trunk and into the roots, trying to build as much strong root growth as possible. So, avoid branch pruning in autumn, especially deciduous trees before leaf fall, until the last of the summer energy has descended to the trunk base and roots. Wait at least a couple of weeks until all the leaves have dropped.

Winter

In late autumn and **early winter**, nearly all the energy is stored in the trunk and down in the roots. It is not only deciduous trees that go dormant in winter; conifers and many broadleaf evergreens also go dormant, even though they keep their green foliage.

So, **late autumn to early winter is a good time to prune branches** because they have little or no sap flowing then.

But don't wait too long, because trees start to "wake up" in late winter. Several weeks before they bud out, the starch in the roots turns back to sugars and the sap starts to surge up the tree. When that happens, you've left it too late to prune off any major branches. As well as the energy being distributed up in the branches, there is also a serious risk of the tree losing large quantities of sap at the prune sites, a mishap that some people call "bleeding". Avoid this by pruning in late autumn or early winter—or wait until late spring.

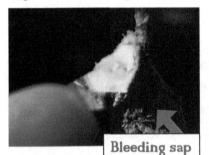
Bleeding sap

Spring

In spring, trees start to bud out and extend stems. As they push out buds, the energy is low in the tree because all the sap is going to the branches all over the tree with the sugars needed to push out new buds. It's a low energy time, so **don't prune your branches in spring**.

Crab apple in spring

Summary:

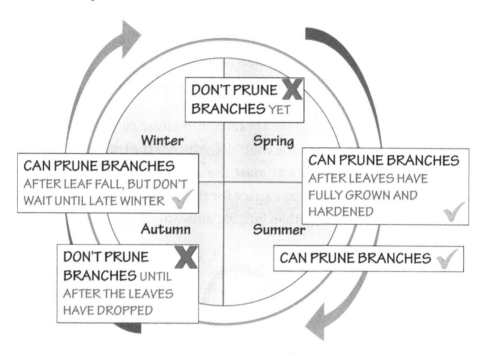

We can prune branches when the tree's energy is high, or medium. That is, late spring through summer, and late autumn to early winter.

In late winter and early spring, we don't do major branch pruning as the tree is pushing out new buds.

In autumn we don't prune branches of deciduous trees before the leaves drop. Wait a couple of weeks after the last leaf has fallen; then it's a good time to prune, or even chop the trunk, if needed.

> **Note:** A few tidy-up snips, and removing suckers from the trunk base, is okay at any time of year. Suckers and long unwanted shoots consume energy as they grow, so remove them before they grow big.

When to prune flowering species

See *When to prune flowering and fruiting trees*, page 107.

Flush cut? or leave a stub?

A "flush cut" is sometimes used to remove a branch from the trunk, leaving the wound flush to the line of the trunk. A curved, concave branch cutter is used to cut into the trunk by a few millimetres (1/8").

In the months after you remove a branch with a flush cut, you get an ugly oval callus growing around the chop site, which is the tree's natural way of healing a wound. So, if you must use a flush cut, a curved concave branch cutter is useful to make space for the scar tissue to callus across the concave wound, rather than bulging outwards.

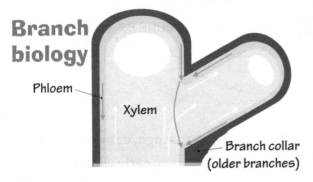

Branch biology
Phloem
Xylem
Branch collar (older branches)

However, is a flush cut healthy for the tree? To answer that, we need to look at the anatomy of a tree where the branch meets the trunk.

The branch collar under older branches contains tissue that helps heal over wounds, and resists entry of pathogens into the phloem, cambium and xylem. When we prune off a branch flush to the parent stem, we remove the branch collar, **eliminating the tree's defences at the wound**. That's not good.

Flush cut

- Okay for new shoots and young trees in development

- Risky for older trees

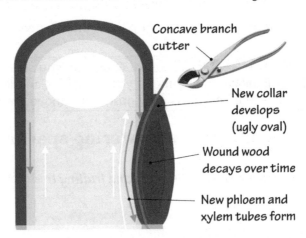

Concave branch cutter
New collar develops (ugly oval)
Wound wood decays over time
New phloem and xylem tubes form

Part 2—Bonsai Reference A-Z: Branch pruning: why, when, how | 75

So, a flush cut could allow pathogens to enter the parent stem. We can use cut paste to seal the wound, but the wound is still in contact with the parent stem or trunk, so the trunk could already be compromised.

Stub cut

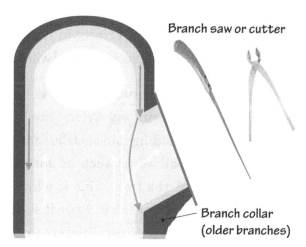

- Keep the branch collar on the trunk or parent stem

- Repairs the wound quicker

Branch saw or cutter

Branch collar (older branches)

When we leave a stub, we retain the stem's branch collar, so there is less possibility of pathogens entering the main stem.

All chop wounds can cause the stem to die back by a distance. In a stub, the dead wood is quickly compartmentalized, and the wound is repaired quickly. However, a flush cut can cause "die back" in part of the trunk.

Advantages of the stub cut:

- Wounds heal faster on stubs, and are less risk to the tree.
- It's relatively simple to make a stub look like a natural feature, whereas the callus around a flush cut clearly appears human-made.
- On conifers, keep a long section of the unwanted branch, and remove the bark. Use pliers to rip back some wood to appear like gravity has torn the branch downwards.

In summary, when pruning a bonsai that you value, it's safer to leave a stub then tidy it up the following year. However, you can safely flush cut "suckers", or thin new shoots without causing harm to the trunk.

Cut paste

All trees produce sap, which is arguably the best wound sealant, containing the natural ingredients to seal and protect pruning cuts. For this reason, when sap flows freely in the growing season, cut paste is unnecessary; trees have evolved to seal wounds effectively on their own.

However, when **winter pruning**, there is little or no sap flow, and the tree's cambium, phloem and xylem, are left open to "the elements". Those elements, including rain splashes and run-off water from nearby trees, can bring pathogens such as fungal spores directly into the vascular system of the tree. This is why we should use cut paste in winter, to protect the organism against such pathogens. Note that cut paste for coniferous trees has a different formula from that available for broadleaf trees, since conifers produce resin as well as sap.

Some bonsai artists use cut paste to heal chop wounds more quickly. However, I have not observed a difference in callusing speed. The main advantage of more expensive putty-like sealant over glue-like paste, is that putty products are easier to remove when appropriate.

Putty wound sealant

Pruning methodologies
Clip and grow

"Clip and grow" is a natural bonsai pruning methodology, or rather, philosophy, where we allow the whole tree to grow out for a few months, then prune it back. We use selective pruning to remove unwanted branches that do not favour the shape of the tree, and to enhance the branch structure over time. I call it a philosophy because in reality, all healthy trees grow, and we need to clip them back to develop and maintain them as bonsai. In other words, **we always clip and we grow** our bonsai in all **pruning techniques**; however, the difference here is that we rely on this philosophy as being the main, or only, method for shaping the tree and developing its branching.

Clip-and-grow pruning works well for trees that grow vigorously and branch rapidly and profusely. A few examples are: elm, olive, sageretia.

Nigel Saunders, of *The Bonsai Zone*, famously uses the clip and grow methodology, and he doesn't use wiring in his bonsai videos. Nigel also uses directional pruning to achieve natural looking results over time.

Directional pruning

If you look at a branch shoot, under each leaf petiole there is a latent bud that can someday get activated and grow into a new secondary stem. As bonsai artists, we use this to our advantage, because after pruning the end of a shoot, the next bud to develop into a sub-stem will be the next one down from the site of your pruning cut.

For example, if you want the branch to continue developing towards the left, snip the stem just above the leaf that points towards the left. Leave sufficient stem length above the leaf so that your prune cut does not kill the axial bud. As a guide for how much length to leave, use at least **double the stem thickness** as your minimum safety margin, and there is no penalty for leaving more... as long as it's beneath the next bud up.

In nature, trees and shrubs can be categorized into those with alternating leaf growth, those with opposing leaf growth, and those with whorl growth. This can make a subtle difference to our directional pruning, depending on the species of tree we're working on.

Alternating leaf growth

Trees with alternating leaf growth also exhibit alternating branch growth. This makes sense when you consider the positions of the axial buds. Elms are a common example of alternating leaf growth.

Opposite leaf growth

Maples are a good example of a species with opposite leaves. When you prune a stem, opposite leaves send new shoots both ways. This is good for bifurcation of branches. For directional pruning you need to decide if you want your next two "child" branches to grow horizontally, (away from the trunk), or more vertically, upwards and downwards. The leaves next to your cut point are either above and below the stem, or at each side of the stem. It's more usual to cut back to a leaf pair at the sides, that will extend shoots horizontally.

Whorl leaf growth

Some trees and shrubs create multiple leaves at one node, and therefore multiple buds, and eventually multiple branches emerging from one point in the stem. An example of a shrub with whorl leaf growth is azalea. Because we want to grow our bonsai with bifurcating branch junctions, when pruning we select only the buds or shoots to leave two shoots emerging from any node or junction, removing all of the others.

Pines produce multiple buds at a node, which grow to candles. But which buds to remove, and which to keep? Consider the relative strengths of the buds. Leave stronger, bigger buds to promote more vigorous shoots, or remove the stronger buds to encourage less vigorous branches. Near the apex, remove stronger buds to leave a weaker pair; conversely on lower branches leave a stronger pair. If you're not wiring the branches, leave two buds on the side that you want the new branches to grow.

Sacrifice branches

Sacrifice branches are shoots grown long in order to thicken or strengthen a specific part of the tree. For example, we sometimes grow them out from the lower trunk over a whole growing season to thicken the lower trunk and increase taper. Then we prune them off in winter, or even the following summer, when they've done their job.

The longer you leave a sacrifice branch to grow, the more it contributes to trunk taper, but also the worse the scar when you eventually prune it off. You need to judge that balance for each sacrifice branch.

Pruning and wiring

This is exactly as it sounds: prune branches to the desired length, and wire them into a position that fulfils the desired shape. Wiring can be helical – spiralling along the length of a branch – or guy wires, anchored to the pot or to another part of the tree. See *Wiring branches*, page 189. If you use wiring, it is likely that you'll also use clip-and-grow directional pruning on the same tree; they are not mutually exclusive.

Structural pruning versus profile trimming

When you consider the overall silhouette of your tree, it is possible to prune only shoots that emerge outside the desired shape. This is sometimes called "profile trimming", or "hedge pruning". It is the quickest and easiest way to prune a bonsai, and is especially useful for a broom-style tree. It is a good pruning method in these circumstances:

- If the branch structure is already well established.
- The bonsai is super vigorous and grows unruly very quickly.
- For a quick summer prune to shape, without major branch cuts.

The hedge pruning methodology has been famously endorsed by Walter Pall. He has developed some magnificent examples of large, naturalistic bonsai that genuinely look like fully-grown mature trees.

However, if you need to improve the primary and secondary branch structure, a profile trim alone won't do the job. To ensure there is bifurcation throughout the tree, and to remove vertical and crossing branches, you need to study each branch, remove unwanted branches, and choose cuts to balance the vigour throughout the tree.

In my day-to-day bonsai work, most of the time pruning is spent on structural pruning, and only a fraction of the time is profile trimming.

Creating an apex

Dome-shaped Apex

The apex is the crown at the top of a tree, and it helps to shape the overall "feel" of a bonsai. It is part of the bonsai's style. A large, dome-shaped apex can make a bonsai appear mature, majestic and venerable. Equally, a sparse, thin apex of a formal-upright conifer can appear naturally tamed by the elements. See *Bonsai styles and styling*, page 50.

The apex is almost the same as a branch, but with two differences: It is supported from below, and there can be two, three or four stems emerging from the same place on the trunk where the apex starts.

To create a large, majestic dome-shaped apex is technically simple; you need only **aluminium wire, a lot of time, and patience**. And three or four thin branches emerging from the same height on the trunk, where you want the apex to start.

Select thin branches to start your apex. Wire them horizontally, and radially away from the trunk. You can use guy wiring or spiral wiring—see *Wiring branches*, page 189.

Top-down view

Tree front

One of the branches might have been a potential new main leader for the trunk; instead of extending the trunk upwards you're stopping it there and bending that down. Bend it directly towards the front view.

Over the following years, secondary and tertiary branches grow in haphazard directions, and some of them will grow right over the top of the apex. That's good: it is gradually forming your tree's crown.

Prune the apex profile to a dome shape frequently using the profile-trimming method. You'll start seeing good results within 2 to 3 years.

Cuttings

See *Propagating by taking a cutting*, page 124.

Defoliation

Defoliation is the process of removing all the leaves from a tree or plant.

Don't defoliate a bonsai unless you know exactly what you're aiming to achieve. Defoliation weakens the vigour of a tree, for two reasons:

- Photosynthesis stops as soon as the "solar panels" are removed. No new energy is generated until the next flush of leaves has grown.
- A trunkload of stored energy is consumed in the process of budding and growing out the replacement leaves.

> **Note:** Never defoliate a coniferous evergreen tree. It won't survive.

Why defoliate?

One reason for defoliating a tree is to intentionally reduce its energy, to reduce its vigour and slow its growth. New stem growth is temporarily halted, and the trunk girth stops thickening. For this reason, defoliation should only be performed on a refined bonsai which you don't want to continue growing taller. Don't defoliate a tree in its development stages.

Defoliating a tree temporarily reduces its need for water. Without the leaves to carry out transpiration, the stems draw up little water. You can lay off the fertilizer for a few weeks, and temporarily reduce watering.

Another reason to defoliate, cited by some bonsai enthusiasts, is to achieve a smaller leaf size with the resultant flush of new leaves. This result is possible, but not guaranteed. The final size of hardened leaves is affected by how much light is available while new leaves unfurl and harden; less light produces larger leaves. So, a valid reason to defoliate the first flush of leaves could be if they grew oversized in gloomy spring conditions. In that case, defoliate in early summer, so that the next flush of leaves is smaller with many hours of sunlight.

But, as already mentioned, only do this to a refined bonsai in order to decrease its vigour, temporarily stop stem growth, and reduce its need for water. And keep in mind:

Defoliating a tree to achieve smaller leaves also results in a weaker tree.

For that reason, it's not recommended to defoliate in the same year as root pruning, since both procedures can significantly weaken a tree.

Some characteristically vigorous deciduous candidates for defoliation are maple, hornbeam, zelkova, and elm. Also, some tropical species such as ficus are so vigorous that they can be defoliated yearly. However, in many cases it's safer and more effective to do some branch pruning to maintain the overall silhouette, and so that photosynthesis continues.

If you must defoliate a tree, don't pull off the leaf and petiole. Use scissors to cut the petiole half way. This protects the new latent leaf bud that's tucked under each petiole. The contrary, removing the whole petiole, could damage that bud and prevent a new leaf from growing.

Developing a bonsai

There are several stages in the development of a plant into a refined bonsai, and it takes many years. Each stage is a subject in itself; as such, this section is a reference index to each relevant section in the book.

Many bonsai artists don't put their tree into a bonsai pot until the final stage—refining the bonsai. However, you are the artist here; you are entitled to plant your miniature tree in a small pot as soon as you'd like.

Keep a log of your work on each tree if you are developing several.

Growing from seed or a cutting

Growing a tree from a seed can take many years before the tree is large enough to develop as a bonsai. It's quicker to take a cutting from a tree that you like; you effectively skip forward 1-2 years of slow growth. Or, **take an air layer** from a thicker stem and skip 5-10 years of development.

See *Propagating trees*, page 123.

Growth, trunk thickening and taper

Unrestricted "full speed" growth is good for a tree's health, and helps to thicken up the trunk. See *Trunk thickness*, page 172. **Fertilize strongly** during this phase, to ensure the tree grows healthily and doesn't lack essential elements. See *Fertilizers—recommendations*, page 98.

Start planning the **style for your bonsai**. See *Bonsai styles*, page 50. If you decide to **wire the trunk** to give it movement to start styling it, do it early in the thickening phase. See *Wiring the trunk*, page 196.

Start developing trunk taper, by growing large sacrifice branches low down, and later performing trunk chops. See *Trunk taper*, page 174.

Performing a major trunk chop is a crucial part of bonsai development, but how do you decide when to chop? See *Chopping the trunk*, page 173.

Training the roots

When the lower trunk girth is thick enough for the bonsai that you have in mind, you need to start training the roots to fit in a small pot. The goal here is to work towards a **near-flat radial root system**. See:

- *Nebari—the surface roots*, page 115.
- *Development and training pots*, page 120.
- *Ten reasons to repot your bonsai*, page 129.

Developing the primary branch structure

This phase is usually done in parallel with training the roots.

Choose thin shoots for your primary branches, on the outside of trunk bends if possible. Remove all the other branches. See *Branch Pruning: why, when, how*, page 65. To help your branch-chopping decisions, first **use a white towel or paper to hide possible unwanted branches**. This technique is famously encouraged by Peter Chan in his pruning videos.

Wire the branches, if necessary, to give them shape and movement that is consistent with the trunk. See *Wiring branches*, page 189.

Prune the branches to start bifurcation and to encourage ramification.

Refining the bonsai

Refining a bonsai largely consists of beautifying it, and establishing the right conditions for long-term maintenance. In particular:

- **Encouraging finer ramification, growing space** and **foliage pads**. See *Ten reasons to prune your bonsai, reasons 3 to 5* on page 66.
- **Creating or improving the apex.** See *Creating an apex*, page 80.
- **Improving the nebari.** See *Nebari—the surface roots*, page 115.

You may have heard the expression:

> *"A bonsai is never finished."*

That's true, because refinement continues into long-term ongoing maintenance, and of course, healthy trees never stop growing.

Fertilizer science

For bonsai lovers, this is a science class not to be missed. The fertilizer recommendations in the next section are based on the science in this section, supported by decades of horticultural testing and evidence.

Firstly, let's dispel some common myths about "plant food":

- Fertilizer is not "plant food", and when you add fertilizer you are not "feeding" your tree. True plant food is photosynthate: the sugars created by the plant itself in photosynthesis.
- Fertilizers are not "nutrients" like proteins. Fertilizers are chemical elements that need water in order to enter plants.
- Fertilizers are not "vitamins for plants". Vitamins have no benefit for plants. Don't bother with "Vitamin B1 for bonsai".

Instead, think of fertilizers as being supplements for plants; essential elements to assist with biological processes, and to form part of the plant materials such as chlorophyll and the growth regulators. For example, cytokinin contains a large proportion of nitrogen.

Tree composition

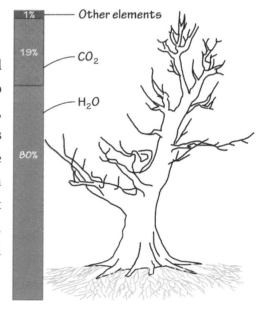

About 99% of a plant is composed of water and carbon dioxide, so every tree is made of 99% carbon, hydrogen and oxygen. This means that only 1% is made up of the other necessary elements. Carbon dioxide is the only chemical that enters a plant from the air. Everything else must be dissolved in water in order to enter a plant.

These essential elements are available in varying amounts in field soil in the wild, and are needed by all plants. Trees in nature can grow their roots long and deep in order to access damp, fertile soil. But we must provide these chemicals for our trees in small pots with inert, infertile soil. A bonsai won't die if we forget to fertilize the soil for a few months, but it may slowly wither and die over a few years if it is never fertilized.

N, P, K, and "micronutrients"

Nitrogen (N), phosphorus (P) and potassium (K) are needed by plants as they grow. So too are thirteen other elements in diminishing amounts: calcium, sulphur, magnesium, manganese, molybdenum, chlorine, copper, iron, boron, zinc, nickel, cobalt, silicon. Some people call the extra elements "micronutrients". As they are not nutrients, a better name would be "trace elements". Each of the 16 elements is just as necessary for plants as each other, but in very different amounts.

These elements don't naturally exist in bonsai soil; hence we need to add them—all of them. If an element is missing, it may difficult to detect at first, but the plant will gradually stop growing and die.

N-P-K numbers

COMPOSICIÓN:
ABONO CE - Solución de abono NPK 7-5-7 con Micronutrientes.

Fertilizer products show three numbers for N, P and K. The numbers are the percentages of each element by its concentration in the product. For example, NPK 7-5-7 means that 7% is nitrogen-based molecules, 5% phosphorus-based, and 7% potassium. In this example, 19% of the product (7+5+7) is composed of N, P, and K, leaving 81% to other components: mostly inactive ingredients like water, plus the other trace elements.

Good fertilizer products also show the concentrations of the other necessary trace elements. If a product doesn't even mention "micronutrients" or similar wording, choose a different product.

Nitrogen (N)

Nitrogen is the element that is most needed and yet most quickly exhausted in a plant. It is also usually the quickest element to be depleted in the soil, because it doesn't bind to soil particles. For this reason, we must frequently replenish the nitrogen in the soil, either constantly using a slow-release fertilizer, or frequently using fast-release fertilizer. But beware: too much nitrogen is toxic to the plant.

Lower levels of nitrogen compared to the other elements, can lead to longer, thinner apical shoots at the top of a tree.

Long side shoots

Higher levels of nitrogen can reduce top-leader apical dominance, and **promote longer side shoots**. Adding more nitrogen does not make a tree grow faster, taller, or thicker. However, it can decrease the uptake of potassium from the soil, and can decrease root growth. It can also make plants more susceptible to pathogens.

Phosphorous (P)

Phosphorous promotes apical stem growth, and is the element most related to plant size. Using fertilizer with **more phosphates can result in taller plants**, assuming everything else is in balance and healthy.

Long apical stem

When a bonsai is in its refinement stage, you can moderately reduce the phosphorus levels to get more compact growth.

Potassium (K)

Potassium is vital to the growth and health of plants. It is used in most parts of a tree. **Adequate levels of potassium** contribute to strong root development, and efficient transport of all the essential elements. Because potassium binds to particles of soil, it can persist in bonsai soil for longer than nitrogen and phosphorus.

Potassium deficiency can weaken a tree's overall health, resulting in poor root growth, impaired water and element uptake, and reduced water pressure in cells, observable as soft, limp stems and leaf petioles.

Poor root development

Contrary to popular belief, more potassium does not improve winter hardiness or frost protection; nor does it make flowers and fruit grow. So there is no benefit in adding more potassium than is optimum, and at the same time, it is detrimental to use less than adequate potassium.

Internode length

None of N, P and K is responsible for the length of internodes in stems. That is controlled by gibberellin growth regulators, which are made by the plant itself, using only Carbon (C), Hydrogen (H) and Oxygen (O). See *Growth regulators*, page 37.

Recommendation: N-P-K levels

I have explained some of the science of N, P, and K in plants, and how adjusting their levels affect your bonsai. The best recommendation is:

> **Use a balanced fertilizer** over a long period of time, and don't try boosting or reducing different elements. For example, use a **6-6-6 universal fertilizer**, that also contains the **trace elements**, and follow the instructions on the packaging.

The optimum amount of fertilizer

Fertilizer manufacturers want you to buy and use as much of their products as possible. Too much fast-release fertilizer can be both harmful and yet wasteful, since much of it leaches out the pot. Too little can stunt a tree's growth.

Wild olive, very healthy

Fertilizers do not speed up a plant's growth more than is naturally possible by generating photosynthates using light and carbon dioxide from the air. When a plant has sufficient quantities of the essential elements, its overall growth is governed by the growth regulators, and by environmental factors such as the amount of photosynthates in the tree. On the other hand, insufficient quantities of N, P, or K can slow down the growth.

> *Fertilizers do not force a plant to grow faster than nature can drive. Rather, a lack of fertilizer elements can stunt growth.*

With that in mind, use *only just* the amount of fertilizer that your trees need—not more, and sometimes less.

Read the instructions on the fertilizer packaging, both for the amount and frequency of application. Be aware that most refined bonsai can use slightly less, and the tree will continue to grow healthily.

Fertilizer is a bad response to plant problems

Don't apply fertilizer as a response to a plant problem that you observe in your bonsai. Typical issues such as suddenly yellowing leaves, unsightly spots, leaf curl, dropping leaves, are **very rarely** caused by a deficiency in one more of the fertilizer elements.

Spider mite damage

Chlorosis means a lack of chlorophyl in the leaves, so they start to turn yellow. It could be due to a deficiency of nitrogen, or a lack of iron, or magnesium, manganese, or zinc. But it is **more likely** to be caused by something different like insects, root overcrowding, or dehydration and drought response.

If one specific element is lacking, you're flying blind—how do you know exactly which one is missing? You could send a soil sample and a leaf sample to a horticulture laboratory for analysis. That's unfeasible for most enthusiasts. You could try a foliar fertilizer with just one element to test if that is deficient. It probably isn't.

In general, **adding more fertilizer does not solve most plant problems**.

More issues in plants are caused by adding too many chemicals than too few.

Over-fertilizing a bonsai

A high concentration of fast-release fertilizer can damage the roots if it is not dissolved in enough water. Some people call this "root burn".

Dehydration and leaf curl

Leaf curl, yellowing and sudden leaf drop in a matter of days can be signs of over-fertilized soil.

Too much fertilizer in the soil makes a tree throw off its older leaves in the same way that a tree responds to drought. That's because the over-fertilized soil dehydrates the tree. Too much fertilizer creates a high concentration of chemicals in the soil—higher than in the roots.

The root wall is a membrane that allows water in, including any dissolved molecules, via a process called osmosis.

"Osmosis"
→ water and dissolved molecules can enter through a membrane

With too many salts in the soil, the osmosis stops. Roots don't take in any water that you pour into the soil, thus the tree dehydrates.

That situation remains until you water the soil enough to flush out the excess chemicals—until the concentration of salts is higher in the roots than in the soil.

Which type of fertilizer should you use?

There are many ways to give your bonsai the essential elements it needs. The main questions that often pop up are:

- Should I use organic or non-organic (synthetic) fertilizer, or both?
- Should I use fast-release or slow-release fertilizer, or both?
- Should I use foliar spray fertilizers as well as soil fertilizers?

We'll answer these questions here. If you're in a hurry, the answers are summarized in the next section *Fertilizers—recommendations*, page 98.

Organic or synthetic fertilizer?

Organic fertilizer

Organic fertilizers come in many different forms. It could be garden compost, chunks of composted animal manure, chicken manure, bone meal, fish emulsion, seaweed extracts, worm castings; the list goes on.

Remember that all of the essential elements that enter a tree (apart from carbon dioxide) must be dissolved in water before they can enter. It is not "healthier to use organic because it's more natural". All dissolved molecules are chemicals, so, whether you use organic or synthetic fertilizer, they're all chemicals entering your tree.

All of these organic products can take several weeks to release the chemicals that plants need. In order to break down the biological compounds and release the chemicals into the wet soil, there must be "good" bacteria and fungi already in the soil.

Here are five **reasons to not choose organic fertilizers**:

Chicken poop in a basket

- You don't know for sure if it contains all of the elements that a plant needs.
- You don't know with certainty how much of the N, P and K you are supplying.
- Most organic fertilizers smell bad for many days after you water them. Definitely not good for indoors!
- You have to wait for the elements to be released by the biological processes in the soil before they are available for uptake in the roots.
- Finally, many insects lay their eggs in faecal matter. They use the chicken poop in your bonsai pots, and after a few days, the larvae start crawling in your bonsai soil. This gets even worse when the birds start to attack your soil in search of those little grubs.

Birds attacked the soil looking for grubs

So, feel free to use organic fertilizers if you don't care about all of the above. But here's my advice:

 Use a synthetic fertilizer, commercially-sold, because you know exactly how much of each element you are applying; you don't inadvertently omit "micronutrients", and you don't attract pests in your bonsai pots. And it doesn't smell.

Slow-release or fast-release fertilizer?

Slow-release fertilizers can be organic or synthetic. When they are synthetic, they are usually called controlled-release fertilizers.

Controlled-release fertilizers (CRF)

CRF: rubber-coated spheres

In the book, *Modern Bonsai Practice: 501 Principles of Good Bonsai Horticulture*, Larry Morton strongly recommends CRF, which comes in the form of small balls of synthesized fertilizer elements, coated with a thin, porous layer of rubber, which slowly releases the ingredients into the surrounding damp soil over several months.

Some advantages of CRF are:

- When the soil temperature is higher, more fertilizer is released.
- The fertilizer chemicals released into the wet soil are complete, and in good proportions for the needs of most plants. They are released in low quantity but in constant supply. You won't "burn the roots", and you also won't get chemical deficiencies.
- You don't need to apply it frequently. Depending on the brand and its instructions, most CRFs can be applied twice a year. By contrast organic fertilizers are needed six times a year, or more.

A disadvantage of CRF, similarly for organics, is that it can be several weeks before you observe a difference in your plant's growth. When you do see improvements, you can't fully credit the fertilizer, because many other factors also affect a tree's growth over a period of months.

My recommendation if you have too many trees, or not enough time to keep fertilizing on a frequent schedule:

> **For 40 or more trees, use a controlled release fertilizer.**
> Use a reputable brand. Sprinkle a good quantity of the little coated spheres onto your soil surface **in spring and in autumn**, following the instructions on the packaging.

If the little green spheres appear ugly, sprinkle top dressing over them.

Fast-release fertilizers

Fast-release fertilizers can come in solid or liquid form. Solids must be dissolved in water; liquids must be diluted. In either case, follow the instructions on the packaging, to ensure you know exactly how much fertilizer you are using.

Watering with fast-release fert

Some advantages:

- You know exactly how much N, P, and K and other elements you are applying to your bonsai soil.
- Diluted chemicals don't attract pests in the soil.

Disadvantages:

- You need to fertilize on a routine.
- When you water your bonsai, it's important to let some water drain out. This means that, whenever you water, some of your fertilizer is leaching out the pot—an unknown amount.

My recommendation if you have not too many bonsai, let's say fewer than 40 trees, and you can fit regular fertilizing into your schedule:

> **For a few trees, use fast-release water-soluble fertilizer.**
> Dilute per instructions on the packaging.
> Make it a fortnightly **routine from spring through autumn**.
> Stop fertilizing if temperatures go above 35°C (95°F).

If you use a **liquid fertilizer**, first check that the soil is already damp, so that you know the tree has already received water before you fill the soil with diluted chemicals.

Bonsai trees don't need a special "bonsai fertilizer"; that costs more for the same product in a small bottle. A balanced NPK 6-6-6 universal fertilizer is sufficient for most bonsai. Check that the packaging specifies the elements that plants need, including "micronutrients".

A balanced NPK 12-12-12 fertilizer has double the concentration of chemicals; you just need to dilute that with double the water compared to a 6-6-6 fertilizer. **Avoid** over-fertilizing and "burning the roots".

Dilute with more water and fertilize more frequently?

Imagine a fertilizer packaging instructs to dilute X amount of product with Y water, and apply every two weeks. You could consider doubling the dilution (twice the amount of water) so that the concentration is half the instructed strength, then using it weekly instead of biweekly. Thus, it's like **drip feeding with less fertilizer but more frequently**.

This is not recommended by the manufacturers because, they say, at half strength, we can't know for sure if enough of every element will get in contact with the roots. But **I have found this works perfectly well** and has no ill effects, as long as you are okay with a more frequent routine.

Foliar-spray fertilizer?

In agriculture, foliar fertilization is used as a rapid, short-term response to a specific deficiency detected in plants. Do you know exactly what is deficient in your bonsai?

It's possible to "green up" a tree's leaves within a matter of days using a spray, but **only if the yellowing appearance is caused by a deficiency in iron, or nitrogen or magnesium**. And such deficiency is rarely the cause of leaves turning yellow on a bonsai.

If there appears to be no other cause for yellowing leaves (like drought, insects, other pathogens, root damage, or stem damage) you could use iron chelate foliar spray as a diagnosis aid to see if iron is missing. If the leaves go green again, then you've diagnosed the issue. But the elements you spray onto the leaves don't have a lasting effect, so you still need to treat the root cause, and fertilize the soil routinely.

> **Don't bother with foliar fertilizers**, unless you're testing for the deficiency of one specific element. Foliar fertilization **won't fix a long-term chemical imbalance** in the tree but might help you diagnose the nature of the problem.

Should we combine multiple fertilizer types?

It is better practice to use one method for a whole growing season.

Here's why: we know that fertilizers provide the essential supplements in various amounts, but we don't know exactly how much of each is getting taken in by the roots. If there is an issue, or if you suspect one type is not as effective as another, you won't know which one is the culprit if you use two different types in the same soil.

Also, using multiple fertilizers increases the concentration of salts in soil. So, unless you frequently test your soil, you won't know if it has too high a concentration, or an imbalance between the fertilizer elements.

> If you want to try multiple methods of fertilizing, use **one type of fertilizer for some of your trees, and another type for others**. Then you'll have a clearer comparison of the methods.

Fertilizers—recommendations

If you have questions about any of these recommendations, or if any seem counterintuitive, see *Fertilizer science*, page 85. Firstly:

> Don't fertilize your bonsai as a response to a plant problem you observe. Nor straight after repotting. First let it recover.

Fertilizer N-P-K levels

Avoid boosting or reducing the levels of nitrogen, phosphorus and potassium in your bonsai trees. Overdoing it could damage the roots, or dehydrate your tree, or "starve" it of another essential element, and you might not observe the damage until weeks or months later.

> Use a **balanced fertilizer** over a long period of time. For example, use a 6-6-6 universal fertilizer, or 20-20-20. Be sure that it also contains "micronutrients" or trace elements.

How much and how often to fertilize

> Follow the instructions on the product packaging, both for amount of fertilizer to use and for the frequency.

For example, a liquid 6-6-6 fertilizer is diluted 5ml product to 1 litre of water (a teaspoon for 2 pints of water), and used every 2 weeks.

When to fertilize

For detailed advice, see *When to fertilize*, page 100. Here's the summary:

> **Tropical trees** – fertilize every two weeks, all year round.

> **Deciduous** and **evergreens (non-tropical)** – fertilize from spring through to autumn.

Which type of fertilizer to use

Organic or synthetic

 Use a **commercially-sold synthetic fertilizer**, because you know exactly how much of each element you are applying; you don't inadvertently omit "micronutrients", and you don't introduce insects in your bonsai pots. And it doesn't smell.

Slow-release or fast-release

Fast-release water-soluble fertilizer:

 If you have **not too many bonsai**, let's say fewer than 40 trees, and you can fit regular fertilizing into your watering schedule, **use fast-release water-soluble fertilizer**.

Slow-release fertilizer:

 If you have **many more trees**, or not enough time to keep fertilizing on a frequent schedule, **use a good brand of controlled release fertilizer**. Cover the small fertilizer pellets with a layer of fine top soil.

Foliar-spray fertilizer

 Don't bother with foliar fertilizers, unless you're testing for the deficiency of one specific element. Foliar fertilization won't fix a long-term chemical imbalance in the tree but might help you diagnose the nature of the problem.

Combine multiple fertilizer types?

 If you are eager to use multiple methods of fertilizing, try **one type for some trees, and another type for others**. Then you'll have a clearer comparison of the different methods.

When to fertilize

Tropical species and indoor bonsai

Fertilize tropical bonsai fortnightly all year round, since most species don't go dormant in winter. In cold climates, keep the bonsai indoors; most tropical species thrive in warm environments, and continue to push out new growth throughout winter. For details on positioning a bonsai indoors or outdoors, See *Guidance by climate*, page 28.

Deciduous, conifers and non-tropical evergreens

Deciduous trees have a shorter growing season than evergreens, so **fertilize deciduous trees adequately during the entire growing season**.

Most evergreens also go dormant in winter, when the foliage turns to dark green/brown, so there's no need to fertilize then. Let the foliage be your guide: when the evergreen foliage turns back to brighter green, and **you see new growth, start fertilizing** again.

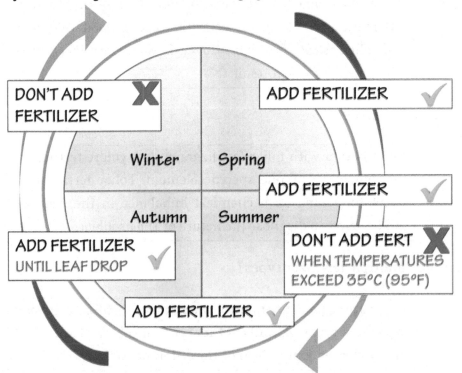

Advice by season

Spring and Autumn

Canopy and stem growth are most active from spring to early summer, and the fastest root growth occurs in early autumn. During these times, ensure that a supply of fertilizer is available to the roots. Many people apply slow-release fertilizer twice a year: in early spring then in early autumn.

Winter and Summer

In winter, deciduous and coniferous trees go dormant, and so they don't need any fertilizer in the coldest period. Wait until you see the first flush of new leaves grow out in spring. Plants don't need fertilizer to start budding out—they manage that for themselves using photosynthates stored in their roots and trunk base. Although, there's no harm if there is any slow-release fertilizer remaining in the soil from autumn.

In summer, as long as temperatures are not stiflingly hot, continue fertilizing to the same regime as in spring and autumn.

However, during the hottest period, many trees go into a state of non-growth, that some people call semi-dormancy. It is not dormant; however, the growth can slow to a standstill while the tree transpires at its maximum rate in order to keep cool. Virtually all the water going up the xylem is used for transpiration, rather than photosynthesis. Don't add fertilizer to the soil during that time, because the additional chemicals are not needed for growth. (Any slow-release fertilizer remaining from spring won't do any harm). But definitely don't over-fertilize, because that reduces the plant's ability to take in enough water.

In the hot summer months, just be sure to keep watering enough!

Flowers and fruit

This may be obvious to you, but needs to be said: Fruits start life as flowers, and all fruits carry the seeds of their plant.

Flowers are the sexual organs of a plant. They offer up the DNA of the plant, in pollen, for nature to help mix the gene pool of that species. Procreation with slight genetic variations "weeds out" the weak, and thus gradually strengthens the species. Flowers are attractive in colours and scents, so that insects in the vicinity visit and pollinate those plants. Flowers evolved to be attractive to insects, because any flowers that failed to attract them would have died out centuries ago.

But not all flowers are attractive to humans.

If you don't want the flowers or the fruit

When a plant grows flowers, and later transforms them to fruits, it takes a substantial amount of the plant's energy, depending on the size and quantity of the flowers and fruit. Flowering trees do not extend their roots while they are developing flowers.

For bonsai, this means we need to be selective about whether we keep the flowers, or remove them. For example, this is a simple choice for **Boxwood**. As a bonsai, box can be very beautiful, but the flowers and fruits are definitely not so.

Some of the most commonly used boxwood varieties, such as Japanese box (buxus microphylla), grow notoriously slowly. This is good for keeping a bonsai small, but it means that we want to direct as much energy as possible to leaf and stem growth, rather than to the flowers and fruit.

Japanese boxwood

When the flowers and fruits are unwanted, **pinch off all the flowers as soon as you see them, to conserve energy**.

Boxwood flower buds

Use tweezers to remove the buds, especially if you have fat fingers like mine.

Gently twist each bud downwards, against the direction of growth. Don't pull!

For tropical bonsai such as **Carmona**, **Serissa** and **Sageretia**, pinch off all the flowers when indoors, to avoid attracting all the tiny insects.

If you want the flowers but not the fruit

You probably know that **Satsuki Azaleas** are famous for their spectacular display of flowers—in terms of quantity, colour, shape and size. There are literally hundreds of different varieties.

They make wonderful bonsai, partly due to the small leaf size but mainly for their beautiful annual floral display.

Satsuki Azalea in June

However, the azalea fruits, in the huge numbers we're talking about per plant, take away a disproportionately high amount of energy from the growth of leaves, stems and roots.

> This means, to keep our bonsai's energy high, that we must **pinch off all the azalea flowers at their base as soon as the floral display starts to wither.** We remove every flower's ovary and its receptacle, so that it does not grow into a fruit.

| Azalea flower | Pinch off the flower using your finger and thumb. | Ensure to remove the seed pod that held the flower. |

If you have a Satsuki bonsai, make this a yearly practice. Don't rush it; set aside an hour or two and enjoy it! I do mine every June, before the last flowers have withered, to avoid any of the seed pods from taking away energy. In fact, some bonsai cultivators remove all the flower buds even before they bloom one year in every three, to keep the stem growth and root growth strong. Likewise, for cultivars that flower twice a year, I pinch off the second flush of flowers so that all the remaining energy goes to root growth, and ultimately tree health.

Cherry tree varieties are also widely cultivated as bonsai because of their beautiful flowers. Some cherry species flower in late winter before they even leaf out. Flowers that last for only a few days do not take a massive amount of energy from the tree, so, enjoy the flowers while they last!

Flowering trees are usually developed in the informal-upright style, sometimes as cascades, and occasionally as broom-style trees.

If you want to grow the fruit

Many bonsai create fruits that are as pretty as the flowers. **Crab apples** make fantastic bonsai, because although the flowers only last a few days, the bunches of resultant fruit can stay on the tree for the rest of the year, as long as you can protect them from the wind and the birds.

Crab apple

The fruits grow in bunches of miniature apples, so for a bonsai, they can be visualised as tiny trees with tiny apples. Beautiful.

Both wind and bird are brutal bonsai enemies where I live, so I bring my crab apple trees indoors for just the week or two that they're flowering. But guess what: there are no bees or wasps in my house, so the flowers don't get pollinated. The flowers don't develop fruit unless we intervene.

 Therefore, to aid the process, **lightly stroke the stamens of each flower using a small brush**, to simulate an insect and to encourage pollination, so that the flower receptacle continues its growth to transform into a fruit.

Brushing the stamens

Of course, if the tree is outside, it gets pollinated naturally by insects so there's no need to do this.

Dwarf pomegranates

Pomegranates create wonderful fruits that can hang on the branches for over a year.

For small- to medium-sized bonsai, I recommend the dwarf pomegranate cultivar, because these produce tiny leaves, and small fruits.

I also have a fully grown pomegranate tree in my garden and have propagated it to develop a bonsai. It is currently still a pre-bonsai, in need of a few more years of thickening. This is the common large-leaf, large-fruit variety, and the fruits can grow up to nearly a kilogram (2 lbs), so it is inappropriate for anything smaller than a large bonsai or a full-size tree. The pre-bonsai gives beautiful flowers, but the fruits are too big and energy-hungry.

> For pre-bonsai pomegranates, **remove the flower and receptacle immediately after flowering** so that the developing fruits don't take water and nutrients away from the developing trunk, branches and leaves.

Plants that produce **berries** can be beautiful as bonsai, for example, **pyracantha** with its tiny orange bunches, or **cotoneaster** with the red berries dotted about like tiny red lights.

Hawthorns in the wilderness produce white flowers and gorgeous red berries; however, it is rare and special to see hawthorn berries on a bonsai. This is because all its energy is channelled to extending shoots and roots until it reaches maturity, which can take up to 20 years. Hawthorn bonsai are more commonly appreciated for their pretty leaves and their beautiful craggy bark.

Styling a fruiting tree

Most fruiting trees in bonsai are styled as **informal uprights**. They can sometimes be developed as cascades to help show off the wonderful flowers; but it's unusual to see a cascade with fruit.

There is a new bonsai style called "**orchard style**", and it is somewhere between informal upright and broom style. A mature apple tree in an orchard has primary branches that emerge from one point around the trunk, above which they form a kind of vase shape. This is a common way to optimize fruit collection.

Orchard

This is how I develop my dwarf pomegranate: although the base has a typical informal S-bend, the branches all fork out in a vase shape so that the fruit hangs down like baubles from branches, to mimic fruit trees in an orchard. See *Bonsai styles*, page 51.

Orchard style

When to prune flowering and fruiting trees

For most trees that flower in spring, prune branches in early summer, after the flowering season is over. This is the perfect time for Azaleas. Pruning them in winter or spring could remove the flower buds.

Alternatively, if you want your tree to continue flowering and fruiting throughout summer, for example with pomegranates, don't prune them until winter. Let them grow, and enjoy the spectacle. Because pomegranates flower right through summer, prune them in winter while they're dormant, because the flower buds have not yet formed then. Note that plentiful flowers and fruits can slow down root growth.

Front of tree & planting angle

When you plant a tree in a bonsai pot, there are some important aspects that significantly influence the appearance of the planting—aside from the choice of the bonsai pot. These include the tree's position in the pot as well as its orientation. In summary:

3 dimensions of orientation (adjusting the tree's angle and tilt): twisting around the trunk base, tilting left/right, tilting forwards/backwards.

3 dimensions of position (adjusting the tree's position): height within the pot, left/right position, forward/backward position.

When you repot a tree, before adding soil, hold the tree approximately in place in the empty bonsai pot, then twist, tilt and move it, to visualize its best orientation and position. See *Potting up*, page 143.

Front view

The front view of the tree is probably the most important aspect to determine. Look for a perspective that shows off the trunk movement, bark, and primary branches. There could be a gap in the canopy like an open window to view the trunk. Typically, there are no branches pointing directly towards the front view; if there are, these "eye pokers" are usually pruned off or wired to one side. There are exceptions, for example in the broom style, as described in *Bonsai styles*, on page 51.

Another aspect that helps choose a front is the surface roots: the best front is often where the trunk flare and roots appear most attractive.

Planting angle

The tree's angle of tilt is an important factor in the bonsai design and style. Experiment with different tilts until you find the angle that best suits the tree and feels good to you.

> **Note:** formal upright and broom styles are directly vertical, i.e. no tilt.

Position in the pot

When discussing the tree's planting position, let's first distinguish between a bonsai pot and a training pot. A tree in development, in a training pot, can be planted centrally, or wherever is most convenient.

However, in a bonsai pot, trees are rarely positioned in the exact centre. Instead, they are often placed slightly nearer to the rear of the pot, offering the viewer more appreciation of the planting and the pot itself.

Similarly, trees can be positioned slightly to the left or right of centre. This asymmetry can contribute to the visual balance of the planting. For example, if the tree features larger branches on the left side, positioning it slightly to the right can help achieve a harmonious balance.

In conclusion, potting a bonsai requires us to visualize all six dimensions – orientation and position – either subconsciously, or by physically experimenting with different positions and tilts before piling on the soil. While some trees have an obvious front and tilt, for most there's no "right answer". Explore various positions and perspectives and ultimately choose what looks best to you. After all, you're going to be admiring your bonsai in its pot for the next 1-3 years.

Fruiting trees

See *Flowers and fruit*, page 102.

Growing a bonsai from seed

See *Propagating trees*, page 123.

Insect infestations, infections

See *Reviving a dying bonsai*, page 154.

Moss

Beautiful, lush moss: it's pretty, has health benefits for plants, and has a water retention capacity that is nothing short of extraordinary. But is moss good for your bonsai? Of course it is! Or is it?

In this section, we're going to explore whether the moss is in fact beneficial for a living tree in a small container, or if it is actually jeopardizing the health of your Bonsai.

Moss growing naturally on the soil surface

Moss myths

Let's start by dismantling a couple of myths about moss.

Myth: Moss generates humic acid, which helps root growth.

This theory falls apart when we consider how humic acid is created. Decaying plant matter creates humic substances as part of the molecular decomposition, and some cultivators indeed add humic acid into soil to assist the uptake of fertilizer elements.

However, moss takes several years to decompose, due to the antioxidants that it generates and stores. This means it will be at least three years, and probably more, before moss starts decaying and creating humic acid. And by then, we have most likely removed the thick mat of moss, and probably repotted too before that happens. **So...**

...**Myth busted!** However, moss can help root growth in another way. Because moss holds moisture better than most soils, and with a dark, moist cover, it makes lucrative haven for tree roots in search of extra growing space. That's the reason why roots grow upwards into the moss.

Myth: The reason that moss doesn't decompose easily is because it contains a lot of lignin.

Moss indeed takes several years to decompose. However, this is not due to lignin, because almost no species of moss contain lignin. Lignin transports water in the veins of plants and trees, and as its name implies, lignifies those veins to wood. Moss does not have veins.

Another myth debunked! The simple explanation for why moss doesn't decay quickly on the compost heap, is because moss contains natural antioxidants which preserve its cellular structure for longer than most other plant matter.

Moss: friend or foe

Having dispelled the myths, let's look at the horticultural benefits of allowing moss to grow on your bonsai soil. After the advantages, we also consider the potential risks, so you know it's not all sunshine and roses.

Advantages of moss:

1. **Superior water retention**: Moss can absorb and retain more than twenty times its own weight in water.

    This amazing capacity to hold so much water means that there will always be moisture at the soil surface, so all those fine roots just beneath the surface won't dry and die in the heat of summer... assuming you don't forget to water it.

2. **Uniform water distribution**: Moss on top of the soil helps to balance the water distribution vertically throughout the pot. In watered soil we get a "perched water table" at the base, because of gravity, and from there upwards the water density becomes less and less saturated, so there is always a vertical gradient of saturation. Since moss absorbs and retains more water than the soil, it wicks up some of the water, helping to balance the water distribution vertically. This is a similar effect to that produced by smaller soil granules at the surface. See *Soil layers and stratification*, page 167.

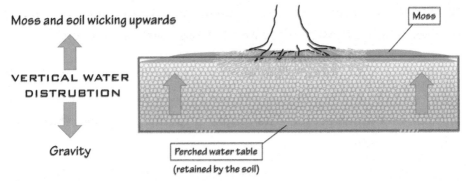

3. **Extra space for root growth**: Trees need space to grow, and we're restricting their root growth by confining them to a small container. Moss can provide an additional half inch of growing space, which is especially useful during autumn, when the roots continue to grow and fill themselves with photosynthates ready for winter storage.

4. **Antiseptic properties**: Certain moss varieties such as sphagnum possess antiseptic properties. It is an ideal medium for packing into an air layer or a ground layer to keep it moist, yet free of pathogens like bacteria and fungi. See *Propagating by air layering*, page 126.

5. **Aesthetic Appeal**: Perhaps the biggest benefit of using moss is that it looks pretty! It looks especially great on "penjing" landscapes, almost as if it were grass in the scene. See *Landscape*, page 60.

Also, if you show any of your trees in an exhibition, you have to cover the soil with moss anyway, and it can add extra interest if there's more than one type of moss or more textures in your scene.

Disadvantages and risks of moss:

1. **Concealing insects**: Moss can harbour unwanted insects and their larvae, and in the worst case, these can feed on tender roots. The vine weevil is one example of such a hostile insect; in autumn and winter its grubs feed on the roots, causing our bonsai to wilt or die.

2. **Attracting birds**: Birds sometimes disturb the soil while hunting for insects and grubs in the moss, especially if you use organic fertilizer. That can leave a mess in your bonsai garden, but more seriously it can expose a lot of roots, potentially damaging the tree if those roots get dry. You can attach mesh on top of the moss to protect it from the birds, but unfortunately that spoils the appearance of the moss.

3. **Water repellence**: Despite its capacity for absorbing a lot of water, in fact dry moss repels water. This means a lot of water can run off the surface and out of the sides of the pot, rather than through the soil. So moss may prevent your soil from getting a proper watering. You can circumvent this by misting the moss before watering, or by watering twice. But without moss, you just don't need to do that.

4. **Obstructing visibility**: Moss on the surface prevents you from seeing the moisture level of the soil. The growing medium might be drying and you could be neglecting it, or alternatively you could be watering when it's already wet.

Too much moss obscures soil surface

5. **Trunk dampness**: A common issue with moss is that it grows on the trunk base, and keeps the bark permanently damp. This can prevent the bark from looking old and cracked up around the trunk base, like a mature old tree. Whenever you see this, pick the moss off the trunk, and wash off any algae using soapy water and a toothbrush.

To conclude, opinions differ. Some bonsai professionals meticulously remove all signs of moss they see growing on the soil because of the risks and issues it brings. Other notable bonsai artists say that the benefits in tree health, like the extra root growth space in moss, are well worth the inconveniences. As practices vary at the highest level, we can be assured that there is no right or wrong here.

As for me, I don't deliberately grow moss on my bonsai soil; it just tends to regrow each year. There are probably moss fragments or spores lying dormant in my reused soil. I leave moss to grow on the soil unless it gets too thick, then I remove it, because that's when the birds attack it.

Tip: Watch out for false moss like **pearl wort and liver wort. Remove these as soon as you see them.** Like all weeds, they are difficult to eradicate once they take hold. They grow roots into the soil, competing with the tree for water. And because they don't retain moisture, they don't uphold the water balance.

If, after reading all the risks of using moss, you want to go ahead and propagate moss to use on your bonsai soil, here is a simple method:

How to propagate moss

1. Lift a small clump of moss from the soil or the shaded side of a tree.
2. Place it in a plant propagator, with a few drops of water to ensure constant humidity.
3. Place the propagator indoors in a bright window, so that it receives daylight, but not in direct sunlight. Moss does not need to be watered as long as there is humidity in the air, so every few days check that there is always a little water inside the propagator.

Damp moss in a propagator

4. Wait 6-12 months and see how that moss has grown!

Nebari—the surface roots

Buttressed surface roots on an old Ficus macrophylla in Seville, Spain

The attractive woody surface roots between the trunk and the soil are known in the bonsai community as "nebari".

These nebari start life as radial roots growing from the trunk, and as they mature and thicken in contact with the air, they develop bark. And just like the trunk, surface roots grow cambium, and xylem and phloem—the tree's veins. So, when you look at nebari, you're really seeing part of the trunk, rather than roots.

From an aesthetic point of view, well developed nebari make a bonsai appear more attractive, mature and firmly grounded.

It's also beneficial horticulturally, because the root collar and surface roots store additional starch during winter dormancy.

On a young bonsai, we try to develop roots that grow radially out from the trunk, all around its base, to gradually thicken and become nebari. And at each repot we work towards a **flatter root plane**, so that the roots fit in a shallow bonsai pot. See *Ten reasons to repot your bonsai*, page 129.

Over time, the flat root plane causes the trunk to develop a root collar— the basal **trunk flare**.

Trunk flare and root flare

As the root collar thickens, it creates accentuated taper at the base of the trunk; a feature that many bonsai enthusiasts refer to as "root flare". In strict horticultural terminology, the correct term for this is **"trunk flare"**.

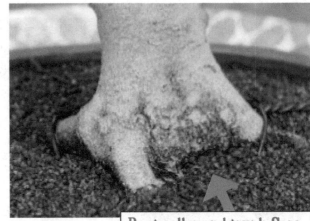

Root collar and trunk flare

However, both terms are commonly used in the bonsai community.

Like the nebari, trunk flare can greatly enhance a bonsai's appearance, giving it a sense of age and maturity. It's especially attractive when the trunk flare slopes into nebari, as one naturally blends into the other. Basal trunk flare also exaggerates trunk taper. See *Trunk taper*, page 174.

To develop trunk flare, as with nebari, we encourage surface roots to emerge radially outwards at the trunk base, all around the perimeter.

If you plant the tree on a flat surface, such as a ceramic tile or the base of a bonsai pot, radial roots will grow out horizontally, perpendicular to the trunk. Horizontal nebari help to develop trunk flare; however, over time they grow disproportionately thick and long, inconsistent with a realistic image of a tree, so you'll eventually need to chop some of the visible surface roots shorter if they grow horizontally.

To appear more natural, and develop trunk flare in harmony with the rest of the tree, the radial roots should slope diagonally downwards from the trunk into the soil, rather than horizontally. Cultivating sloping roots develops a more a natural and realistic transition from trunk to nebari. We can encourage sloping radial roots using a simple trick when repotting. It's called the "stone trick". Read on...

The stone trick: develop great nebari

If your bonsai has a relatively thin trunk, plant it on top of a rounded stone, inside the pot, so that roots cannot grow directly downwards.

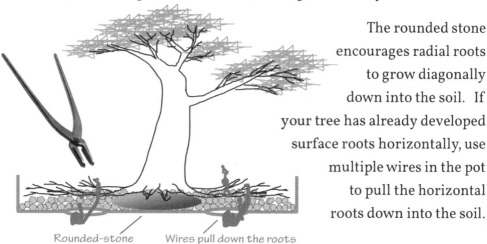

The rounded stone encourages radial roots to grow diagonally down into the soil. If your tree has already developed surface roots horizontally, use multiple wires in the pot to pull the horizontal roots down into the soil.

Rounded-stone   Wires pull down the roots

If the trunk is very thick, plant it on top of a downward-facing ceramic plate. Do this at each repotting for several years, and you will gradually develop an attractive root collar with realistic-looking surface roots

Upside-down ceramic plate

that slope naturally from the trunk to the ground. Keep the developing root collar covered with soil for a few repots (which could be between four and ten years); don't leave the emerging surface roots uncovered until you see they have thickened to a size consistent with the trunk girth. Once you repot leaving the surface roots exposed, they more-or-less stop thickening. Or rather, they thicken in proportion with the rest of the trunk—which is exactly what we want.

Pests and pathogens

See *Reviving a dying bonsai*, page 154.

Pot choice

Which do you choose for your bonsai—a shallow pot or a deeper pot? Ceramic or plastic? Glazed or unglazed? What shape? Read on…

Aesthetic aspects

An entire book could be written about the magic union between a bonsai and its pot. The nuances are not the focus of this guide; however, here are some **basic rules of thumb for a refined bonsai** or tree in refinement:

General guide

Firstly, consider the idea of "**visual mass**", which is the approximate bulk density you perceive of a tree or a group planting. This is not scientific; it's just an approximation based on your artistic perception. In the excellent book, *Principles of Bonsai Design*, David De Groot suggests that:

> *The visual mass of the pot should be about a quarter that of the tree (foliage and trunk).*

Note: Some artists might consider the following list too simplistic for such a beautiful and intricate art; however, everyone needs to start with basic rules before understanding how to vary or break the rules!

Pot size:

- Pot width—about two thirds the tree height.
- Pot depth—about the same as the trunk diameter near the base.
- Exception for cascade style—pot height is visually proportional to cascading tree height. Usually more than quarter the visual mass.

Pot shape (viewed from above):

- Rectangular—for robust trees with thick trunk and imposing apex.
- Oval—for a curvacious trunk line, slender trunk or sparse canopy.
- Circular—for tall trees such as literati style, or slanting style.
- Square or hexagonal—for cascade and semi-cascade styles.

Material/finish:

- Glazed ceramic—deciduous trees and large-leaf evergreens.
- Unglazed—coniferous trees and some small-leaf evergreens.
- Colour—match some aspect of the tree. For example: bark colour, leaf colour in spring or autumn, or flower colour, etc.
- If in doubt, unglazed brown/grey pots suit more-or-less any tree.

Lip and lines:

- Exterior lip around the pot—for more imposing, bulky trees.
- Inward curve up to pot mouth—for slender trees with subtle details.
- Horizontal line around the pot—the pot appears shallower than its true height, which can be useful for a thin tree with a deep root ball.

Horticultural aspects of pot size

In simple terms, the bigger the pot, the more soil space your roots have to grow into. However, it's not necessarily correct to say "The bigger the pot, the healthier the bonsai". This is because pot depth plays a major part in water retention in the pot, and the vertical water distribution.

In every absorbent material we get a "**perched water table**". In a pot, this is a completely saturated layer of soil, and won't drain out the drainage holes. It's held there by the balance between gravity pulling it down, and the capillary action in the soil trying to wick up the water.

The perched water table is **caused only by the soil's water retention**, and not by the pot floor. You can test that this is true by looking at the soil in the base of a pond basket or colander after watering: once the excess water is drained, the soil at the base is still completely saturated.

This means, for a given soil type and granule size, that the perched water table is the same height, regardless of the depth of the pot. So, just after watering, a shallow pot will be more waterlogged than a deeper pot.

Now let's consider what this means for the roots:

The perched water table gradually evaporates off or gets absorbed into the roots. But just after you've watered, the roots in a shallow pot are going to be much more waterlogged for a while. For more details on the perched water table, see *Soil layers and stratification*, page 167.

If you choose a shallow pot, you don't want the roots to permanently sit in waterlogged soil, starved of oxygen. Therefore:

> Mound up the soil and plant your tree a bit higher, or use a prop stone, as detailed in *The stone trick* on page 117.

Here's another tip for cooler climates and colder times of year:

> After watering, tip up the shallow pot to let water drain out the lower end. That'll get oxygen to the soil and to the roots.

Development and training pots

Deep grow pots

If you want to thicken up the trunk of your bonsai, you could either put it in a bigger, deeper pot, or plant it in the ground. This way, you give the roots more space to grow into, and leave it for three to five years. You wait for it to grow bigger, and don't prune the canopy or the roots.

But when you have a bigger container, the top half is going to get much drier than in a shallower container. So, the roots are encouraged to grow naturally downwards, deeper in search of more water.

Japanese maple in a deep grow pot

Therefore this is only suitable as a growth phase for a pre-bonsai, because after this, we need to train the roots for a shallower bonsai pot.

Ceramic or terracotta training pots

Many pre-bonsai are delivered in Tokoname clay or terracotta pots. They're a great way to cultivate a healthy bonsai during its refinement and before finally potting in a small ceramic bonsai pot. For cheap training pots, they are not unattractive.

Tokoname clay training pot

Colanders or pond baskets

Similarly to clay training pots, a pond basket can be used as a pot in the development of a pre-bonsai, to encourage a mass of fine roots which are best for the health of a tree. The pot retains the granular soil, and the soil retains the water.

Pond basket

Because there is air around all the edges of the soil, when a root grows to the edge it stops extending in the dry air. The root tip sends a chemical signal back along the root, to branch out new fine roots nearer the trunk.

So, a pond basket avoids the development of long, winding roots, and encourages fine, radial roots of an even thickness. It's an excellent way to develop a healthy root system before repotting to a small bonsai pot.

A tree can stay in a pond basket for several years. There's no need to prune roots until you determine it's time to transfer to a ceramic pot.

In contrast to a clay training pot, a pond basket is decidedly unattractive. It is a horticultural system, clearly not suitable for displaying a bonsai.

Pot refresh

If the roots don't need pruning significantly, it is a healthy practice in spring to perform a pot refresh. This is a cursory soil cleanup and root check, using the same pot as before. No wiring is required, because the cleaned up root ball fits perfectly into the old pot.

How do you know when to do a pot refresh instead of fully repotting the bonsai? See *Five reasons not to repot*, page 138.

To do a pot refresh:

1. Use a root rake or spatula to remove moss from the top soil. It is often necessary to remove a thin layer of compacted top soil too.
2. Use an old toothbrush to remove moss and algae from the surface roots and trunk.
3. Gently lift the soil mass out of the pot. If there are no winding roots around the periphery of the soil, put it straight back in the pot, and skip ahead to step 5.

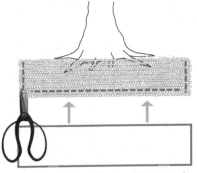

4. If you see winding roots around the soil, prune them off only around the edge and under the soil mass.

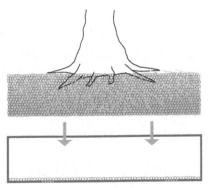

5. Put a thin layer of granules of fresh soil in the pot base.
6. Replace the root ball in the pot, and add some fresh top soil.

Aftercare: No special aftercare is needed following a pot refresh.

Propagating trees

The three main ways to propagate a tree are by planting their **seeds**, taking a **cutting**, and taking an **air layer**.

Propagating by seeds

It is slowly, deeply satisfying to create a bonsai from a seed. However, be aware that it takes many years before the tree grows large enough to develop as a bonsai—cuttings are quicker, and air layers quicker still.

If you decide to grow from seeds, here are some tips:

- Use at least ten seeds, because not all seeds germinate and grow.
- Place the seeds in a bowl of water for fifteen minutes. Reject any that still float, because they are not viable and will not germinate.
- Research the species because some seeds must be cold stratified over winter to simulate their native winter conditions before they germinate in spring. Most species don't need cold stratifying.
- Acorns are an example of a seed that doesn't need stratification. To germinate them, wrap each one in a damp kitchen tissue, then seal them in a plastic bag. Put the bag in the fridge (optionally) to reduce the risk of mould.

An acorn germinated in wet tissue

- Check the seeds weekly. Pot each seed as soon as its root emerges. Point the root downwards in the soil. Use vermiculite or similar to maximize root growth in the first year. See *Soil components*, page 164.
- Place the pots in a plant propagator box to keep the humidity high.
- If it's winter, place the box on a thermostatic heat mat to keep the temperature at about 24°C (75°F) and position a grow lamp above.

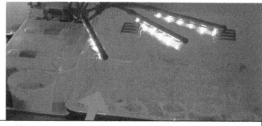
Seedlings in propagator with grow lamps

Propagating by taking a cutting

Propagating a plant by taking cuttings is quicker than growing from seed. You effectively skip forward 1-2 years of slow growth, and you clone all the characteristics of the "donor" tree. Here, we explore ways to maximize each cutting's chance of growing roots in the least time.

 Take several cuttings to increase the overall chance of rooting.

The best time to take cuttings is from **mid spring through early summer**, to maximize sugar production and auxins, both needed to grow roots.

Hardwood and softwood cuttings

Research the species that you're propagating, because some plants generate new roots more easily on softwood cuttings than hardwood. For example, Japanese maple cuttings have a higher success rate of rooting softwood cuttings than hardwood. A softwood cutting has less than one year's growth, and the stem has not yet lignified its xylem.

In a hardwood cutting, the previous year's xylem has lignified, forming a woody interior. This is the process that makes rings annually in the heartwood of a tree. Likewise, the outer dermis thickens to form bark.

Some species root more readily than others; you can take thicker, older cuttings from trees that root more easily. For example, olives and elms root easily from hardwood, even on cuttings up to 1.2cm (1/2") thick.

How many leaves to keep?

Select a shoot with 2 to 10+ leaves. For species with large leaves, keep only 2 or 3; for tiny leaves, keep more than 10. **More leaf surface area makes more photosynthate**, but also increases the water demand and the risk of dying due to insufficient water supply—so you need to strike a balance. **Use a plant propagator** to reduce transpiration in the leaves, to decrease the water demand. Ultimately, keeping more leaves and using a propagator increase the likelihood and the speed of rooting.

Preparing a pot for your cutting

Use a small, clear plastic pot, and poke some tiny holes into the bottom. Clear plastic lets you see when roots reach the edge of the pot. Put in some absorbent soil, like peat moss; you'll get rid of that soil once you see roots. **Wet the soil, then put the plastic pot on a small drip tray to hold a shallow pool of water.** Pooled water may rot roots, but not before any roots exist. Poke a hole into the soil where you'll plant your cutting.

Taking a cutting

1. Choose a cut point just below a bud or leaf node.
2. Use a **sharp, clean cutting tool** to reduce the possibility of infection. Cut cleanly, then immediately place the cutting in water.

Preparing the cutting (hardwood only)

3. Use a sharp, clean knife to cut a line into the bark and cambium around the entire base of the stem, about 1cm (1/2") up from the bottom. Remove the bark beneath your cut line, also removing the green phloem and brown cambium. The remaining stub helps keep the cutting stable, and encourages the roots to grow radially.

Rooting the cutting

4. Dip the stem in **rooting hormone.** See *Rooting hormone*, page 38.
5. Plant it in the hole in your soil, then gently push the soil inwards onto the stem just firmly enough to hold it still.
6. Place the pot and tray in a **plant propagator,** in bright light.
7. **Check the cutting monthly**, to see if it has struck roots. If it has, keep it in there for a few more weeks, to grow more roots.
8. Refresh the pot's water monthly, then weekly after roots appear.
9. When there are sufficient roots, transplant the cutting into a larger grow pot, taking care not to damage the roots. Use a sturdy granular soil that will maintain its structure for a few years. See *Soil*, page 161.

Propagating by air layering

The fastest way to propagate a tree to make a viable bonsai is to take an air layer. This can skip ahead 5-10 years of development time.

An air layer is the removal of bark around a stem, which is then covered with damp sphagnum moss to grow roots while still in situ on the donor plant. Research your species; some plants root more easily than others.

The best time to start an air layer is from **spring through early summer**, to maximize sugar production and auxins, both needed to grow roots.

Prepare your materials

You will need: a marker pen, a sharp knife, rooting hormone, sphagnum moss, clingfilm plastic wrap, aluminium foil, and duct tape. In the absence of sphagnum moss, spongy lawn moss is also very effective.

Choosing the location of the air layer

Select a branch that you envisage can be the trunk of a future bonsai. On that branch, choose a location that appears to naturally flare out or bulge slightly, so that you can use that as the new trunk base. If you want to create a twin-trunk or a triple-trunk bonsai, choose the branch division that you want to become your multi-trunk base. The ideal location for new roots to emerge is just beneath the bulge or the branch split, but not so far beneath that it creates inverse taper.

Creating the air layer

1. Draw a line around the branch at the location that you chose. Draw a second line below the first, a bit further than the branch thickness.
2. Use a clean, sharp knife to cut cleanly, following the lines you have drawn. Push the blade right through the bark as you cut.

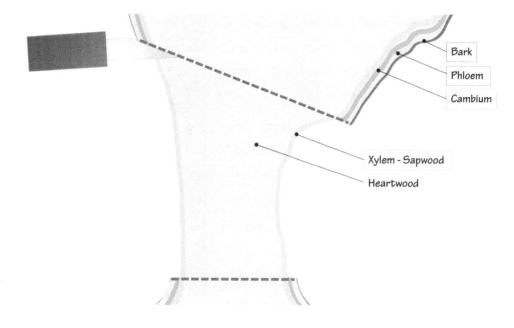

3. Strip off all the bark between your two lines. Strip off all the visible green phloem, and all the brown cambium beneath it.

Removed an inch of bark, phloem and cambium.

Note: Don't leave any cambium, because that would create scar tissue, and could prevent the layer from creating roots.

It's okay if you accidently scrape off some of the xylem (sapwood); however keeping the xylem is less risky for the branch above the air layer

4. Keep the layer damp. Paste rooting hormone powder or gel on the bare sapwood, making sure to completely cover the upper cut line.

5. Wet the sphagnum moss, and plaster it completely on all the bare wood, especially the upper cut line.

6. Wrap the clingfilm or plastic wrap loosely around the layer, and use duct tape to seal the bottom to the bark beneath the layer.

7. Fill the loose plastic with damp moss. Mix in some akadama to facilitate unravelling the roots from the growing medium.

8. Wrap it tighter, and seal the top of the air layer with duct tape.

9. Cover the packet with aluminium foil, to keep it dark and allow roots to grow, yet letting you periodically inspect progress. If the moss is starting to dry, inject water into the packet.

Air layer with plastic wrap and duct tape

10. When you see roots through the plastic wrap, leave it another couple of weeks until the roots have grown prolifically.

Air layer has enough roots

11. Use a branch saw to chop the base of the bare wood, taking care not to damage the roots. Remove the packing and unravel the roots.

12. Pot your new layered tree in a deep pot with granular bonsai soil.

Aftercare: Treat it as a recently repotted tree. Place it in semi-shade for a week or two; keep the soil watered but not continually flooded.

Pruning

Pruning branches – see *Branch pruning: why, when, how*, page 65.
Pruning roots – see *Repotting: how to repot a bonsai*, page 142.

Repotting: why and when

Wherever you are in the world, springtime means one thing for bonsai enthusiasts: it's repotting season.

When you water the soil, if it takes a long time to drain through, there could be an issue with the soil or the roots. It might be time to repot.

After pruning the roots

When we talk about repotting a bonsai, we usually mean **pruning its roots**. Merely moving an intact root ball to a bigger pot is not considered to be truly repotting; we call that *slip potting*, which is explained on page 160.

Ten reasons to repot your bonsai

Now let's get serious: repotting your bonsai poses a certain amount of risk to the organism of the tree, so why do we risk repotting it? Here are five horticultural reasons – that deal with the health of the bonsai – and afterwards, some artistic and practical reasons.

Horticultural reasons

1. In our bonsai, we want to develop a root system that is a **fibrous mass of very healthy, fine roots**. This is because the thin, pale feeder roots are the ones that are most healthy; the ones that absorb the water and dissolved

Healthy fine roots

elements from your soil. By pruning longer, thicker roots, we encourage more fine roots to grow near to the trunk base.

2. We sometimes need to **relieve a very badly root-bound system**. When a bonsai is root bound, also called pot bound, the roots start to compete with each other for water and essential elements; there is nowhere new for them to grow, and the health of the tree starts to decline.

Root bound

3. We sometimes need to refresh or **replace the old compacted soil** that could be starving the roots of oxygen.

Old compacted soil

In soil like that, the old clogged up substrate might be wet, but each time you water, almost none of the fresh oxygenated water reaches the roots.

Your tree has a serious chance of getting root rot in that bad old compacted soil. Therefore, remove the old compacted soil and replace it with fresh granular soil. See *Soil*, page 161.

4. We sometimes need to **remove old, rotten, and dead roots** from the root system. The health under the soil surface is critical. Old, rotten decaying roots can contain harmful bacteria, and you can sometimes smell this on the underside of a bonsai pot. If that's the case, it's time to remove the bad roots and the soil.

5. By pruning the roots, we **stimulate the growth of new root hairs** as the root system recovers.

Rotten roots

This triggers the production of a growth regulator hormone called **cytokinin** in the roots. In low concentrations, cytokinin promotes cell divisions in roots, leading to more root branching and elongation. Biologically, by pruning long roots, we encourage the development of finer roots and more root forking near the trunk, by promoting new root hair growth and cytokinin production.

Cytokinin is also transported up the tree's xylem tubes to the branches, where it promotes bud and leaf growth. Evidently, this hormone is absolutely necessary to keep the whole tree growing.

When cytokinin molecules build up in high concentrations in the roots, they inhibit further root growth. Thus, the same naturally produced hormone can either promote or limit root growth, depending on the amount of it relative to auxins, another type of growth regulator hormone. See *Plant hormones*, page 36.

Aesthetic reasons to repot

6. You might simply want to fit your bonsai into a smaller pot.

Smaller pot necessitates root pruning

It's important to understand that thick woody roots, while helping to stabilize a tree, take up a lot of valuable real estate inside a bonsai pot.

So, by removing the thick, woody roots we make more space for the small fibrous roots, which, as previously mentioned, are best for the health of the bonsai. Take this opportunity to remove the big tap root (right), especially if that is preventing the tree from fitting into a shallow pot.

7. When you repot, **remove circling wrap-around roots** that coil around the inside of the pot, or even wrap around the trunk itself.

Wrap-around root

These are not detrimental to the health, but over time they could mark the trunk base and restrict the growth of the attractive radial surface roots.

8. Use repotting as a way of **improving the radial surface roots** and

Flow from trunk to radial roots

the visual flow from the trunk base into those surface roots.

Developing a radial root system helps to thicken the root collar at the base of the trunk, which over many years gives trunk flare. This helps to exaggerate trunk taper, which is always beneficial at the base of a bonsai trunk, because it helps to convey the impression of a mature, old tree trunk. See *Trunk flare and root flare*, page 116.

It also has health benefits, because the trunk base and the woody surface roots provide winter storage for all the glucose that turns to starch over the dormant period. See *Nebari – the surface roots*, page 115.

Surface roots and trunk flare

9. Repotting is the time to remove ugly surface roots that are either crossing over one another or not growing radially away from the tree trunk. When I say ugly surface roots I'm talking about:

- uneven in their thickness
- uneven in their spacing around the trunk
- obvious root chops
- kinked roots that stick out of the soil like an upturned toe
- knotted roots or crossed roots.

Uneven, crossed roots

None of these are a health-hazard to the tree, but if they are disproportionately large, they can detract from the impression of a real-life tree in the wilderness.

10. Repotting is a good chance to sort out adventitious roots that are growing out from the trunk a short way up above soil level.

Adventitious roots

They are not nebari; they are false roots. They grow there because in the past the underlying root system has been starved of oxygen and the tree's way of resolving that is by sending out roots close to the surface. The problem with these roots is that if we let them thrive and grow bigger, they take valuable resources and send water and nutrients up the trunk through those points. They effectively limit the development of the main trunk below the adventitious joints, so if your tree has these false roots then repotting is a good chance to remove them.

It is also important to understand that these adventitious roots are not air roots like those that grow on a Ficus tree from the branches or from higher up the trunk. No: these are a stress response to not having enough oxygen in the damp soil down below.

> **Caution:** Don't prune off adventitious roots if the main root system has too few fine roots. That means the tree has become dependent on the adventitious roots.
>
> If that's the case, tourniquet the trunk below the adventitious roots; then after a few years remove the lower trunk, to convert the upper roots into the main root system. See *Tourniquet*, page 169.

Did I say there are ten reasons to repot? I was wrong; here's another:

11. Cracked pot. For example, in winter it's possible that ice has cracked the pot, so spring is a good time to move the tree to a new pot.

Ice-cracked pot

If the roots have not completely pervaded the soil, consider doing a simple pot transfer, where you do minimal root pruning and keep the majority of the root ball and soil intact. See *Pot refresh*, page 122.

What if the pot has completely broken, and it's the wrong time of year to repot? For example, in autumn, a strong gust of wind could blow over a top-heavy bonsai off its shelf. In this case, it is safer to slip pot it into a larger container without touching the roots, then the following spring do a complete repot, pruning the roots as appropriate. See *Slip potting*, page 160.

When to repot

I take the horticultural view and repot when there is the least risk to the trees. And that is **in spring**. To understand why this is, we need to look at the energy cycle of a tree throughout the seasonal year.

In winter, most trees are dormant; think of that like a rest period while there's no growth at all. This is true for deciduous, conifers, and most evergreens. During that time, all the energy is stored away in the trunk base and roots as starch, which also acts as a kind of antifreeze to some extent. **It's important NOT to prune the roots during winter** because you'd be effectively cutting off a large energy supply that is needed in spring. This is the energy calendar for root pruning:

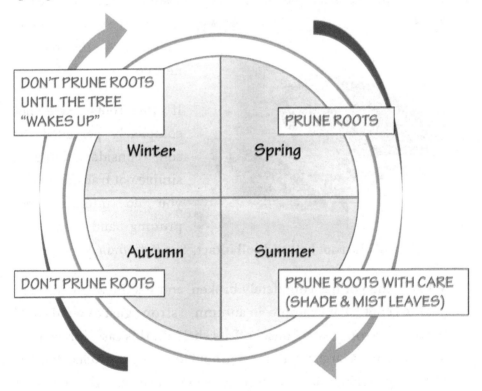

In **later winter or early spring**, as the weather starts getting a bit warmer and the tree starts sensing longer daylight hours, that starch gets converted back to glucose, maltose and sucrose. Dissolved in water, these sugars are what we call the sap, which moves up the tree's phloem in spring and pushes out new buds. Now all that energy is being spent up in the branches, and the tree is still going to bud out regardless of what happens to the roots at this time. Which makes it an ideal time to prune the roots (but a bad time to prune the branches).

In summer the tree is creating copious amounts of new glucose via photosynthesis through the leaves, so summer is a high-energy time for the whole tree. This means if you didn't already prune the roots in spring but you need to then you can still do in summer. But in this case, be cautious of removing so much root mass that the remaining roots cannot supply enough water for the leaves to transpire in the hot, sunny summer days.

Finally, **autumn** is a bad time to prune branches and a bad time to prune roots, because the tree is trying to pack down the last of the energy and store it away for the winter.

Exceptions to the energy calendar

For **tropical species** like Ficus, Sageretia, Carmona, and Serissa, these trees can happily bud and leaf out all year round in a warm environment, so you can prune the roots at any time of year – and you can prune the branches at any time of year. For these species, I recommend you leave a time gap of a few weeks after pruning the branches: wait to prune the roots until you see new leaf buds emerging. A bit like a deciduous tree waking up in spring, all the new leaf buds developing means that plenty of sap is flowing in the phloem tubes up in the branches, which is a "green light" to go ahead and prune the roots.

Five reasons not to repot

It is common characteristic of bonsai beginners to launch headlong into this hobby, with the desire to do everything possible to their tiny trees in the first few months. Repotting is a dearly desirable activity, especially if you have just bought a beautiful new pot for your lovely tree. But first, consider these situations where you should exercise patience and not yet repot.

1. If it's the wrong time of year. As we saw in the energy calendar, there are a few wrong times of year to prune a tree's roots. In autumn, your tree is storing away the last of the energy down into the lower trunk and into the roots, so it is understandably not a good time to be pruning off those roots. In winter, while the tree is dormant, all of that energy is stored in the lower trunk and in the roots, so again, not a good time to be pruning off those roots.

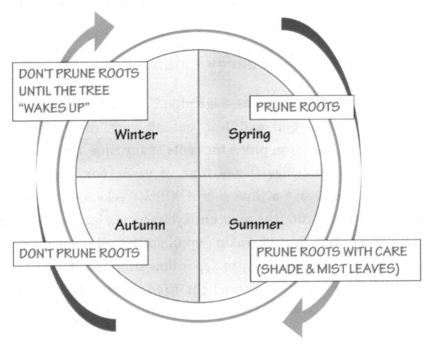

Note: In an emergency, you can slip pot your bonsai into a bigger pot at any time of the year. See *Slip potting*, page 160.

2. If your bonsai is suffering in some way—if your bonsai's health is not great, for example, if it has a fungal infection or an insect infestation. In those cases, you need to treat the fungus, or get rid of the insects, but don't repot it at those times.

Mealy bug infestation

While your tree is recovering from those problems it will likely throw off its old leaves, and it needs all its energy to push out new buds to re-grow those leaves. Once you see plenty of new buds, use that as your cue to repot, assuming it's still the right time of year.

3. If you **want your tree to grow bigger**. If you want to thicken up the trunk or grow the whole tree bigger, just leave the roots undisturbed. Let the roots do their work: let it grow. Pruning the roots of a bonsai stunts the size of the tree.

It is a well-known phenomenon that once you put a tree into a small pot, you're not going to see much growth of trunk girth after that.

I don't expect this little Japanese maple (right) to thicken up much at all in the next ten years realistically, and if I wanted it to, I'd put it into a much bigger pot and not prune the roots.

Small pot, no trunk growth

4. If you're doing some **major branch pruning**. If you are going to cut off some big branches and remove a lot of foliage mass, conserve as much root mass as you can, to keep the energy level high in the tree, to heal the chop wounds and start to push out new growth.

I'm talking about a well-developed, mature bonsai; we need to keep the tree's energy high for its long-term health.

Major branch chop

Some wise bonsai artists have said of their more mature trees: "only perform one major insult per year".

It's not the same as a large, vigorous nursery plant that you are starting to train as a bonsai for the first time. These have masses of roots and foliage, and you can chop top and tail without worrying. See also *Should we balance branches and roots?*, page 147.

And once again, you can slip pot at any time of year into a bigger pot without problem, even if you've pruned off some big branches.

5. The fifth reason not to repot your bonsai and prune the roots is when those roots haven't yet filled out the container.

Roots don't yet fill the pot

If the roots are still growing healthily and there is still space in the container for them to grow further, let them continue to grow in the same pot. It'll be good for the long-term health of the tree.

As roots grow, their microscopic root hairs start to pervade the tiny pores between the granules of soil, or even right into particles of porous substrate. Just by taking out the pot and looking at the roots, we risk disturbing those tiny root hairs; however, we sometimes need to carefully remove the whole root ball from the pot in order to inspect the roots. Try to leave the root ball undisturbed as far as possible when inspecting. Don't remove any soil unless you see that it is in definite need of a pot refresh or root pruning.

In this root inspection (above), I don't need to repot yet because the soil is still looking healthy and it is not pot bound. The roots have not pervaded the whole pot yet, so I'm carefully going to put it straight back in the pot to grow for another year.

As always, exceeding my numbered lists, (did I say there were five reasons?) here's a sixth reason not to repot your bonsai:

6. If you don't have enough fresh bonsai soil. Whatever the soil mix that you're using, if you don't have enough of it, don't start the job. First get the supplies, and then repot. What you shouldn't do is half repot then leave the roots exposed to the air while you go out to get some more soil. The roots may dry and die, so don't start the job if you can't finish it. But there is a handy tip, if you need to leave the tree unearthed for a while. Read on...

Prepare enough soil

Need a break while repotting?

You can leave a root ball standing in a bucket of water for a few hours, which keeps it alive and healthy **as long as**:

- the roots are not showing any signs of root rot,
- it is in the shade, and
- you are comfortable with bare-rooting the tree. See *Should we "bare root" when repotting*, page 146.

Standing in water, mid-repot

Repotting: how to repot a bonsai

Following the energy calendar on page 136, if your bonsai needs to be repotted, do it in spring. At other times of year, if it needs repotting, slip pot it into a bigger pot. See *Slip potting*, page 160.

It's best not to water just before repotting, to avoid wet, sticky soil.

> **Reminder:** when we talk about repotting a bonsai, we mean pruning its roots; not just moving the whole root ball to another pot.

Root pruning

In summary: Unpot, remove old soil, prune roots, tie wire to the new pot and add soil, plant the tree and wire it in. Step by step:

1. Quick check: is the tree still wired down into the old pot? If it is, cut those wires on the underside of the pot.

2. **Ease the root ball and soil out of the pot**. If it is easy to lift, and you see the roots have not pervaded the whole pot, consider waiting another year, or do a rudimentary clean up—see *Pot refresh*, page 122.

   However, if the root ball is jammed hard in the pot, and you need a tool to prize it out, that's a sure sign that the roots need pruning.

3. Prune off the long and winding outer roots, in order to access the soil and the inner root system.

4. **Remove the old soil**. Use a pointed stick or root hook for field soil, or a root rake for granular soil. For now, leave some of the old soil sticking to the roots; it can help them stay slightly damp for a while.

   Pruning off the long and winding roots

5. If you have decided to bare root your tree, wash off all the old soil. You can use a hose with a jet of water, or a bucket of water. I use both. See *Should we "bare root" when repotting*, page 146.

Keep the roots moist. Mist them often if it's warm or windy.

6. Comb out the fine roots radially away from the trunk. Identify roots that have potential to be nebari. See *Nebari-the surface roots*, page 115.

7. **Prune off roots that are**:
   - thick and woody (but don't remove the nebari!)
   - long and winding.
   - pointing downwards.
   - wrapped around the trunk.
   - crossing over other roots.
   - protruding from the trunk at odd angles. Keep only the radial roots.

Keeping the radial roots

8. Trim the roots all the way around the shape of the new pot. Some people call this a "profile trim"; I call it a "bowl cut". Leave the roots slightly shorter than the edges of the pot.

   See also *How much root mass to remove*, page 153.

9. Place the tree in a bowl, in the shade, while you prepare the pot. Spray the roots with water to keep them humid. If you bare rooted the tree, you can stand it in a bowl of water.

Potting up

10. Choose the new pot. See *Pot choice,* page 118.

11. **Prepare the pot**:
    - Put drainage mesh over the drainage holes, wiring each one down with a double-loop of thin wire.
    - Feed an aluminium wire through the holes, ready for tying the tree into the pot. The wire must be long enough for the two ends to protrude several inches above the rim of the pot, so that they can easily cover the surface roots.

- If there's only one drainage hole, loop the wire around a short stick, just longer than the diameter of the hole. Place this under the pot, so that the two wire ends can protrude from the one hole.

12. **Prepare the soil**:
    - Use dry, granular soil. See *Soil*, page 161.
    - Seive all the soil to sift out all the dust and fine particles.
    - Pour into the pot a base layer of good draining soil, like pumice, which won't break down and clog up the drainage holes.
    - Pile on a layer of your main bonsai soil mix. Mound up the soil in the position where you'll place the tree.
    - Alternatively, add a rounded prop stone, so that the roots must grow radially away from the trunk. See *The stone trick*, page 117. Add a thin layer of soil granules on the stone to prevent air gaps.

13. Plant the tree on the soil mound or the stone. Squish the tree down while gently twisting it, to push the bottom roots fully into the soil.

14. Get the tree's angle right—see *Front of tree & planting angle*, page 108.

15. Tie the wires over the surface roots, twisting the ends together. Use pliers to pull the wires upwards and simultaneously twist them.

16. Add more soil.

17. Poke a chop stick into the new soil repeatedly to ensure the soil granules get in between and beneath the roots.

18. Tap the pot gently with your fist or a rubber hammer if it's a big pot.

Poking down the soil

This settles the soil in the container, and removes any air pockets that later become water pools. You may then need to add more soil.

19. Finish up with a top dressing of fine soil granules on the surface to keep vertical water distribution more uniform. Some people use sphagnum moss on the soil surface for this purpose after repotting.

20. Take a photo of your beautifully potted bonsai. **And then water it!**

Aftercare

Repotted triple-trunk elm

After repotting the bonsai, follow these tips for a speedy recovery:

- Place it in bright shade – out of direct sunlight – for a week or so.
- Water it whenever you see the top soil starting to dry.
- Don't add fertilizer for a month or so after repotting.
- For broadleaf evergreens, and for deciduous trees that you have left until summer to repot, consider putting a plastic bag over the entire bonsai for a week, to keep the moisture near 100% restricting the transpiration. See *Revival tips – Tip 4 – plastic bag trick*, page 158.

Repotting mistakes

If, after repotting, you realize that the angle or position is wrong, don't force it around in the soil; you might tear some of the remaining roots.

If you want to adjust the angle, it's essential to weigh the risks and benefits of unearthing and repotting again. Leaving the tree at the wrong angle might complicate future repots, with roots growing at an unwanted angle. On the other hand, two repots in a short time is risky.

If it's the **same day, or even a day later**, very carefully unwire and unearth the tree, then perform the whole potting procedure again. Don't leave it longer, because the root system is starting to develop new miniscule root hairs into the soil, and unearthing could destroy these.

Repotting FAQs

Every spring, in repotting season, I am asked a few similar questions, which probably arise from confusions due to the multiplicity of conflicting information from many sources.

Here I explain the basic horticulture underlying each question, so that you can make your decision for each species, each tree, and each repot. Hence my answers here are not "black and white", yes or no; but explaining the horticultural fundamentals of each question.

Should we "bare root" when repotting?

Bare rooting a tree means removing all of the old soil from the roots.

It is common practice to leave some soil on coniferous trees when pruning their roots. The rationale for leaving on some of the soil is that it contains a large amount of "good" fungal content, called mycorrhizae, which grows in symbiosis with the roots. That is, the roots grow more healthily when they are in contact with mycorrhizae. As well as conifers, several different oak trees also have fungi in the soil, and an accompanying distinctive smell when repotting. So, if we follow this practice of repotting conifers without bare rooting them, then we should also follow the same principles for oaks, and all trees.

However, I have always bare rooted oaks when repotting, with no observable issue to their health. The fungus grows back in the soil again each year, perhaps from spores sent out from the fungus in other oaks' soil during the year.

I have also bare rooted junipers and pines, washing off the old soil in a tub of water. And I have never lost a juniper or a pine after repotting.

So, the question is not about the risk of the tree dying, but more around total root health and speed of recovery after root pruning. For this reason, while I bare root my conifers to clearly see the root spread and structure, I re-introduce a small handful of the old soil, with the fungus if it's visible, into the new soil mix so that any mycorrhizae can continue to grow. My experience has shown that I don't actually need to do this, because the tree will be just fine without it; however, adding in some old soil doesn't actually harm the trees—at least, almost never…

Question: are there any circumstances in which we MUST bare root a tree?

Answer:

Yes, if the roots were rotting in the old soil. In this case, it is imperative to wash off all of the old soil, to avoid transferring the pathogens from the old to the new soil, and reduce the likelihood of further rotting. For the same reasons, prune away all of the rotten roots. While a few drips of stagnant water probably wouldn't infect the new soil (consider how much naturally rotting material there is below the ground!), we can at least try to side-step the possibility of more root rot after repotting.

Should we balance branches and roots?

Every spring, in repotting season, this question comes up in viewers' comments posted to my YouTube videos:

Question: When I prune the roots, should I prune a roughly equivalent amount of branches to balance the roots?

Answer:

It certainly sounds like a reasonable proposition to say that chopping a *roughly estimated* amount of branch material to balance the pruned roots at the same time can reduce or stop the remaining foliage from drying and dropping. Indeed that's the traditional view.

But modern bonsai horticulture views the overall energy level in the tree as more important for its long-term health.

Question: What is the problem with chopping roots and branches at the same time?

Answer:

When we remove a significant amount of roots and branches, we seriously deteriorate the tree's energy reserves. So, to develop and protect the long-term health of the organism, you shouldn't do that unless it is a super-vigorous grower in good health.

In Spring, all the sap – transporting up the stored energy – goes to pushing out new buds and leaves, and flowers and fruit and seeds. So, if you prune significant branches in spring, you are removing a significant amount of energy. And then it can "bleed" sap too.

But pruning roots in spring is not a big deal, because most of the energy is up in the trunk and branches.

When leaves dry and drop, that doesn't normally mean the tree is dying; it means the tree is conserving water by throwing off some of the older leaves. Indeed, this can be to compensate for a depleted root system unable to supply enough water for the transpiration that happens in leaves. But, for most tree species, leaves are disposable; they can be dispensed with temporarily while the tree recovers. And letting the tree decide which leaves to drop is a natural process; it is usually the older and bigger leaves that get thrown off, so it can even help to keep the appearance of a natural tree by keeping small leaf size.

In **conifers**, that could be bad if a whole branch dries. Therefore, for some coniferous species it is necessary to limit your root pruning to perhaps 1/3 of the root mass, or maximum 1/2, and keep at least half the roots in order to avoid the loss of a significant branch. You can do some branch pruning to avoid this risk, as long as you are prepared for this to make a dent in the energy reserves and take longer to return to vigorous growth again. In some juniper species, this can also cause the plant to resort to growing immature needle foliage, perhaps as a natural defence mechanism. When this happens, let it grow unhindered for a year or more until the mature scale foliage regrows. Don't remove the needle growth; that could prolong its state of defence and grow more needles.

In **deciduous** and **tropical trees**, dropping the leaves doesn't cause a branch to die (unless it's for other reasons like a fungal infection). In the majority of cases, dropping the leaves conserves water in the tree stem, by cutting off the transpiration in each leaf. So, in these trees, you could say it is a self-defence mechanism when the tree endures drought.

Question: Is there some way to prune roots that avoids any loss of foliage due to the depleted root system?

Answer:

Yes. In spring, chop the roots but not the branches, then stop the transpiration in the foliage by putting a **clear plastic bag** over the entire tree for a week or so. See *Tip 4 – Plastic bag trick*, page 158. This maintains the humidity at nearly 100%, so that the leaves don't draw up much water for their transpiration.

If you live in a warm climate with many sunlight hours, keep it in a bright but shaded spot for week or two. After a time, the remaining roots grow new root hairs and the tree starts supplying more water again to the leaves.

For deciduous trees in early Spring, there's no need to cover with a plastic bag, even after you prune off a significant proportion of the roots. This is because just as the leaf buds are swelling or starting to unfurl, they are using the sap in the branches for the millions of cell divisions; but they are not yet drawing up a lot of water from the roots.

That'll happen later in the season once the leaves are fully grown and photosynthesizing.

When the foliage is ample or the leaves are very big, that's when the clear-bag trick is most effective. If you leave a repot until early summer, this trick is very useful to keep the leaves green in the otherwise hot, dry air. But, as previously mentioned, it is not a big issue for a deciduous tree to throw off many of its leaves in summer, as long as the tree was previously vigorous and healthy. They will grow back.

> **Note: Don't prune roots in late summer or autumn.** The tree is attempting to pack the energy down into the roots at that time.

> **Anecdote:** In my early bonsai days, I lost a pomegranate after repotting in autumn. The young tree dropped all its leaves to retain moisture, and then started budding out again in mid-autumn. The resultant leaves were small and dropped in mid-winter. It didn't come back to life. I conclude that this was because after root pruning in autumn, the limited remaining energy all went to leafing out again, too late in the growing season to generate photosynthates for winter storage, needed ready for the springtime wake up.

I use the plastic bag trick for broadleaf evergreens – indoor and outdoor. Examples include: azalea, boxwoods, carmona, holm oaks, cork oaks, olives, privets, sageretia, serissa, etc.

Question: Why refute the teachings of traditional bonsai practice?

Answer:

When the traditional practice says "balance" the root pruning with an equivalent amount of branch pruning… how can we possibly know what is balanced? How do we know how much water the foliage or the branch stems draws up, compared to how much root mass is needed to supply that water? Is it ounce-for-ounce roots for branches? I doubt it.

Or cubic inch for cubic inch? Probably not. It's impossible to measure in terms of water supply and demand.

Also, it is certainly different at different times of year. And for different species. There is such a lack of exact science here, that obeying a broad statement to "balance" the tree parts, is at best unpredictable, and at worst is wrong for some species and circumstances, and could jeopardize the long term health.

For these reasons, to prune major branches off deciduous and temperate evergreens, **wait several weeks after root pruning**, at least until you see new leaves have fully grown and hardened.

And vice-versa for tropical evergreens: I first prune the branches in spring, then I wait until I see new buds forming all over the tree before I repot and prune the roots in late-spring.

In general, the best practice is:

> *Prune roots when the sap is rising;*
> *prune branches when new leaves harden*

Notice I wrote "major branch pruning". It's perfectly okay to trim a few minor twigs, leaves, or growing tips to keep the silhouette of the tree in shape at the same time as root pruning. I should also say "major root pruning" for the kind of practice I'm talking about—removing more than half of the root system; sometimes up to 90% of the root mass. See *How much root mass to remove*, page 153. For example, trident maples and crab apples root extremely vigorously because the leaves generate so much energy during the growing season. Even so, it's better to keep them vigorous and healthy, rather than removing all their energy.

Notable exceptions that can be pruned top-and-tail together

Some species are so vigorous that it seems you can do almost anything to them without harming them. It could be considered that significantly reducing the energy in such bonsai is essential to keep them small.

- **Ficus microcarpa** varieties. These are tropical broadleaf evergreens, and don't follow the same energy calendar as deciduous or coniferous trees. It is acceptable to prune branches and roots at the same time. You reduce their energy, but they bounce back. It seems that the only way to kill these trees is to let them get completely dry. Reducing branches and roots seems to have no ill effects.

   > **Note:** Many Ficus bonsai are sold with the name "Ficus retusa", but these are in fact Ficus microcarpa. They make great bonsai.

- **Portulacaria afra (dwarf jade).** These are succulents; not trees. You can chop anything off; branches, roots, leaves, or a trunk chop, then put the remaining plant in dry soil for a week or so, and it will come back to life. It might drop a few dry leaves in the meantime, but that's to retain water in the stem ready to push out new growth.

- **Elm species**, to some extent. Elms are both vigorous and resilient. This is why Chinese elms make such great starter bonsai, as well as their naturally small leaves. It is possible to chop roots and branches together, and see the tree bounce back unscathed within weeks. **But, to be safe**, I treat my elms as the same as any deciduous trees: repot in spring; prune branches in summer and winter.

   > **Note:** Chinese elms are sometimes sold with the name "Zelkova parvifolia", but they are in fact Ulmus parvifolia—a species of elm. Zelkova is a different genus of tree (also an excellent bonsai).

- **Olives**. These trees are also very resilient and with enough light are extremely vigorous. They can survive top-and-tail pruning without problem. Having said that, to be safe I still use the clear-plastic-bag trick in semi-shade for a week or so after major root pruning.

One final exception: removing minimal root mass. If you do this, it's fine to prune the branches at the same time. See *Pot Refresh*, page 122.

Question: Can I use soil without sifting out the fine dust?

Answer: You can, but you're likely to clog up the pores between soil particles and compromise the drainage in your pot sooner than you would do with dust-free soil granules. Sieve it now to save time later; it could potentially add an extra year before the next repotting is necessary. See *Ten reasons to repot your bonsai*, page 129, and *Soil*, page 161.

How much root mass to remove

Question: How much of the root mass can I prune off when repotting? No more than one third?

Answer: Maybe this is a good guide if you're a beginner pruning the roots for the first time – to only remove a maximum of one third of the root mass – to not risk losing the tree. However, it depends on time of year, and on the species, and the general health and vigour of the tree.

In **early spring, deciduous trees** can lose up to 90% of their root mass, and continue to grow without problem.

Most **coniferous trees** are more sensitive to losing root mass. **In spring,** a conifer can lose half its root mass, or up to two thirds as long as the tree is healthy and vigorous.

Most **tropical trees** can lose two thirds of their roots, or more if they are just budding out; then they bounce back without problem. Use the plastic bag trick to avoid leaf loss. See *Tip 4 – Plastic bag trick*, page 158.

The **rest of the year,** don't prune the roots significantly of any tree. If it's becoming rootbound and losing vigour at "the wrong time of year", slip pot it into a large pot, then the following spring do a thorough job of repotting and root pruning. See *Slip potting*, page 160.

Reviving a dying bonsai

If your leaves start turning yellow, or brown, and start to drop, firstly…

DON'T PANIC!

Overreacting could worsen the problem. Of course, if the roots got dry then you need to water the soil very soon. But **don't overwater it**, because leafless trees don't need much water, and we don't want to suffocate the roots. And **don't use fertilizer** until it has recovered because those salts could cause the roots to lose water back into the soil.

Also, let's rule out the obvious: If it's Autumn and your tree is deciduous, its leaves will naturally go yellow or brown ready for winter, in which case you don't have to do anything. Be aware that some trees go yellow a long time before others; it can even happen in late summer.

Now, before attempting to revive your bonsai, first make sure it is actually still alive. The best way to do this is to scratch into the bark with a fingernail on part of the trunk. If you see green cambium, it means it is still alive. But if you only see brown wood then it's unlikely to be recoverable, and probably already dead.

Green cambium layer: the tree is alive

Try it also on another tree of the same species to be sure that the cambium is indeed green on a healthy specimen.

Let's have a quick look at what's happening inside the tree.

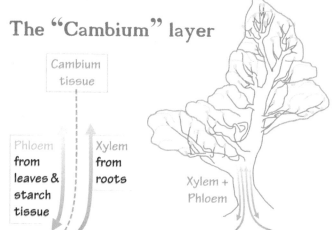

When we talk about the green cambium layer, we really mean the phloem.

Phloem tubes transport the sugars generated in the leaves, and they also contain chlorophyll which is why we see them green. When a section of phloem dies, it loses its green pigmentation because the chlorophyll breaks down to its component elements, thus losing its colour. However, **it is very possible that the cambium is still alive** and the xylem is still able to transport up water, so that being the case, we just need to give the tree some time and the right conditions to bud out again. Read on to see what the right conditions are!

Revival tips

Here I explain each tip to show why it is successful, and when to use it.

Tip 1 — pests and pathogens

Inspect the tree closely for insect infestations or fungal infections, and deal with either or both as soon as you detect them.

Insect infestations

Miniscule insects can decimate a plant, but like a "successful" disease, they rarely kill their host; they just consume its sap and suck away all the vigour. However, **root-eating grubs like vine weevil larvae can kill** a plant by destroying the fine roots. Vine weevils are small black beetles that feed on the leaves, then lay their eggs in the soil.

If you see insects such as **scale, mealybugs or spider mites**, these are all wickedly elusive, and even if you remove all the visible insects, their eggs and larvae are probably still in crevices all over your tree. Mealy bugs and scale are also notoriously difficult to eradicate, because their waxy skins repel water, and therefore repel water-based insecticides.

If you see several **ants marching into the pot, this spells trouble**. Ants farm smaller insects like scale, mealybugs and aphids, by bringing these insects onto your plant in order to consume their sugary excrement. Yes: ants eat the poop of herbivore insects. If there is a sticky substance on the leaves or soil, this is another sign that you have an infestation.

So, if you see the signs of an infestation:

> **Use an oil-based insecticide like neem oil every week for four weeks** to ensure that as eggs hatch, the larvae can't reach maturity and so can't lay any more eggs.

Neem oil is also called "azadirachtin", and despite this name, it is in fact a naturally occurring chemical from neem seeds. It doesn't instantly kill adult bugs, but it fully deals with all the larvae and immature insects. Treat persistent bugs using a cotton bud dipped in the oil, or alcohol.

Fungal infections

If it's a simple fungal infection like mildew:

> Use a plant-friendly fungicide. Spray the tree and soil surface.

If it's a more serious fungal infection like leaf spot or juniper-apple rust:

> Remove all of the infected leaves and carefully dispose of them so that there is no cross-infection to other trees.

Whether you use insecticide or fungicide, be sure to follow the instructions on the packaging. If your bonsai loses more leaves during either treatment, don't worry; we'll catch that with the fourth tip.

Tip 2 — flowers

When flowering trees suffer, they sometimes produce a flush of flowers as a stress response, perhaps in an evolutionary attempt to spread their seeds before dying. However, we want to maximize the chances of recovery, by channelling the tree's limited resources to leaf growth, in order to generate new sugars through photosynthesis. Moreover, flowers attract insects, and this is not a good time to encourage an insect infestation. See *Flowers and fruit*, page 102.

> Pinch all the flower buds, flowers, and seed pods off a suffering bonsai, in order to redirect all the tree's energy to leaf and stem growth. Don't remove the leaves or leaf buds.

Tip 3 — roots

Check if root issues are causing the decline. If you water and it doesn't easily drain through the soil, two things could be wrong: it might be root bound, or the water could be stagnant, slowly rotting the roots.

Gently ease it out of its pot to inspect the roots at the level of the pot floor. If you observe wet, sodden, limp black roots, they are rotting.

> So if it's **springtime, you need to repot**. Trim the long roots, and remove the rotten roots.

See *Repotting: how to repot a bonsai*, page 142.

> In **late summer, or autumn, or winter,** you might need to perform emergency **slip-potting** to avoid disturbing the healthy roots.

See *Slip potting*, page 160. You'll still need to prune off any rotten roots.

Whether root pruning or slip potting, **always use granular soil** for good drainage and strong root growth, ensuring the best conditions for your tree to recover. See *Granular soil*, page 162.

If there is moss on the soil, keep that in your slip pot, because roots could be growing up into the moss. See *Advantages of moss*, page 111.

Tip 4 — plastic bag trick

The now famous "plastic bag trick" is a seemingly magic method of reviving a near-dead bonsai.

 Place a **clear plastic bag over the entire tree** to keep the humidity at nearly 100%, but still allow light to the leaves.

This virtually stops the remaining healthy leaves and soil from losing water through evapotranspiration, and encourages the tree to push out new buds. Spray some water mist into the bag occasionally, and keep the bonsai in a bright place, but not in direct sunlight.

To understand why this works so well, we need to consider the transpiration. See *Photosynthesis and transpiration*, page 33.

Large clear plastic bag

In short, when a leaf photosynthesizes, it releases oxygen and water through the thousands of stomata into the air, and this is called transpiration. But leaves don't only photosynthesize; they are also the cooling system for the tree. Each leaf draws up water through the xylem in the stem, and most of the water evaporates into the air to cool the leaf, and to cool the water and sugar that goes back down the phloem.

So here's the key point: to revive our tree, we need to **greatly reduce the amount of water that the leaves evaporate off by transpiration**, because the roots and xylem might not be able to supply enough water. To do this, we need to keep the humidity as close to 100% as we can, hence the plastic bag with water misted inside. We keep it in a bright place but not in the direct sunlight, so that it doesn't get too hot.

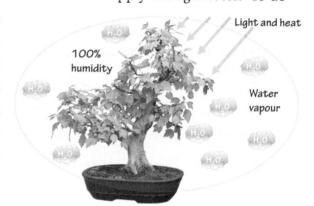

About once a month, spray a small amount of **plant-friendly fungicide** in the bag, to help prevent mould.

Even if you think your bonsai is dead, keep the plastic bag on for several months, watering lightly. You might eventually be happily surprised by a new shoot, even when all hope seems lost.

Summary

Whether you use insecticide, fungicide, repotting, or slip potting, always use the plastic bag trick to revive a bonsai. Patience is key and every tree has its own timeline for recovery, so don't stop watering!

In bonsai, if patience is king, persistence is the queen.

Roots

For **root growth** and **rooting hormone**, see *Roots, water and soil*, page 35.

For **root health** and **root pruning**, see *Repotting: why and when*, page 129.

For **surface roots**, see *Nebari—the surface roots*, page 115.

Slip potting

Slip potting is the simplest and most risk-free way to repot your bonsai. In truth it's not really "repotting", because you don't prune the roots; you just **transplant the tree and root ball into a bigger pot** with fresh soil.

1. Prepare your new, larger pot, with a couple of long wires looped through the drainage holes. Add a base layer of pumice and some fresh bonsai soil. Mound the soil in the middle.
2. Ease the root ball out of the old pot. If it's jammed in hard, use a spatula or a repotting knife around the edge, and lever it out.
3. Use a rake to loosen the tightly wrapped roots, freeing the root ends.
4. Place the root ball into the new pot, and squish it down onto the new soil while twisting the trunk back and forth to thoroughly embed the roots into the new soil.
5. Tie the wires over the root ball.
6. Fill the edges with fresh soil.

Slip pot

Aftercare: no special care is needed after slip potting. Place it back in its previous location, and continue your normal maintenance regime.

Slip potting **should only be used in "emergencies"**, because a couple of years after you slip pot, it can cause some bad tangling of the non-radial long roots. An "emergency" could be a broken pot outside the repotting window in spring, or a badly suffering tree in dire need of a larger pot.

So, if you slip pot one year, then the **next spring, repot it properly** with root pruning, provided the tree is healthy. See *Root pruning*, page 142.

Soil

Bonsai soil often sparks lively debate amongst cultivators. Each enthusiast refines their own blend of soil components to suit their trees, local climate, and watering needs. But it's a simple choice when we consider these **three fundamental requirements** for bonsai soil:

- it must **retain enough moisture** to keep the roots damp,
- it must **allow good drainage** with a granular composition,
- it must **preserve its structural integrity** so that it will still allow good drainage after one or two years, or longer.

So, from the second and third requisites, we can immediately see that peat moss, compost substrates and loamy soil are unsuitable for bonsai.

One more desirable feature of good bonsai soil is an unmistakable **colour change between wet and dry**, so that it is easy to see when your bonsai needs watering.

It's obvious that soil must retain water for your tree's roots, but why is it so important to allow good drainage?

Why drainage is critical

Good drainage

It turns out that roots don't only need water and fertilizer elements; they also need **oxygen**. Each time you see water percolating through the soil surface, it sucks down some air with it. Oxygenated water keeps your roots healthy and happy. The alternative, stagnant wet soil without oxygen, allows harmful microbes to grow and enter the roots, and will start to rot those roots. When this happens, you can smell the sour, putrid odour under a pot. If you smell that, thoroughly flush out the soil with water, and consider slip potting into fresh granular soil.

Granular soil

As a rule, **bonsai soil should be granular**, in order to allow sufficient drainage between the granules. While loamy field soil may retain water for longer than granular soil, its dense, muddy composition restricts the passage of excess water.

Granular soil

Beyond its drainage properties, granular soil has another advantage **during repotting**. Unlike clumpy field soil or compost, which can be difficult to remove before root pruning, **granular soil is easy to remove**. This makes the repotting experience less arduous and more enjoyable.

 Before deploying your soil, sieve it thoroughly to **sift out the fine particles and dust**. Soil dust clogs up the pores between granules, and compromises the drainage in your pot.

Granule size

For any given soil type, a cluster of small particles retains more water than the same size cluster of large granules. This is one reason why the powdery particles of peat-moss retain water so well. However, small particles break down to finer particles and dust, and clog up the pores within soil and the drainage holes. As a result, the very property of retentiveness in small particles can also potentially jeopardize the drainage, accumulating waterlogged, stagnant clumps of soil.

To strike a balance between water retention and drainage, the **optimal size of soil granules is between 4–5 mm (1/6"–1/5")** for most bonsai.

However, for mame-sized trees in tiny pots, a smaller granule size is preferable, between 2–3 mm (1/12"–1/8"), to retain more water than

that retained by larger granules. Particle breakdown and blockage are less of a concern for mames, because we usually need to repot them sooner than large bonsai.

Small soil granules 2-3mm (1/12 - 1/8 inch)

The optimum soil

The ideal soil for your bonsai is likely to have mix of components so that it fulfils all three requirements—water retention, drainage, and structural durability. For example, small granules of lava rock aid drainage and hold their structure for decades, but they don't retain much water. Conversely, akadama retains ample water, but it breaks down to a clay mush after a few years. A blend of these two components makes a decent balance, by both maintaining the structural stability and retaining water.

Soil mix : 50% Akadama, 50% Kiryuzuna

Soil acidity

In general, one soil mix should suffice for most of your bonsai; indoors and out. Most soil components have a **pH level of about 6.5**, which is very slightly acidic, and favourable for root growth of many species.

However, there is an important exception for acid-loving plants, like azaleas and camellias. These plants thrive in soil with a lower pH value, meaning higher acidity. We can achieve this with Kanuma soil: it has good water retention and drainage, maintains its structure adequately, and importantly, **Kanuma is naturally acidic with pH 5.5**.

Soil components

Here is a table of components with comparative ratings for structural durability, water retention, and colour-change behaviour when wet.

> **Note:** These are some soil types commonly used in bonsai horticulture. Other soil types are available.

Soil component	Structural durability	Water retentiveness	Changes colour?
Akadama	2	3	Yes
Cat litter (pink)	3	3	Yes
Expanded fired clay balls	3	3	No
Gravel	4	None	No
Kanuma (acidic)	2	3	Yes
Kiryuzuna	3	2	Yes
Lava rock	4		No
Peat moss, seived	1	4	No
Perlite	3	1	No
Pine bark, chopped	1	4	No
Pumice	3	2	Slightly
Vermiculite	1	4	Slightly

Ten of the twelve above-listed components are inorganic, meaning that they are derived from minerals. The only organic materials here are peat moss and pine bark.

Akadama

Akadama is a solid clay, mined in Japan. The words "aka" and "dama" roughly translate to "red pearls". This is my favourite soil component: it is hard enough to maintain its structure for a couple of years, retains water, and even allows root hairs to grow in its microscopic pores. It shifts colour shade when wet, so it makes an excellent top dressing. One disadvantage: akadama breaks down to clay dust after several freeze-thaw cycles, so it is not the best soil if you live in a very cold region.

Kiryuzuna

"Kiryu", as it is sometimes called, is structurally harder than akadama. Kiryu is similar to pumice, but is slightly denser. An advantage of kiryu over pumice is that it changes from light yellow to orange when wet, so it is a good addition to a top dressing. Like pumice, it retains less water than akadama, so in hot regions it needs to be mixed with more retentive components. Also, unfortunately, kiryuzuna is not cheap.

Cat litter

Certain types of "pink" cat litter are absorbent, structurally stable, and cheap. However, not all cat litter is suitable: avoid the clumping sort, and beware that added scents and chemicals can harm the roots. Before you use a product for your bonsai, test it wet; freeze and thaw it to assess its structural stability. If it leaves a mush, don't use it.

Expanded fired clay

Some calcined (baked) clay aggregates are absorbent, structurally stable, and light. Expanded clay nuggets are usually too big to use as bonsai soil; however, if you find a brand with 6–8mm size granules (1/4"–1/3"), this makes a good base layer in deeper bonsai pots.

Perlite and Pumice

While the minerals in perlite and pumice are different from each other, there are some startling similarities between these soil components. They are both white, both mined from volcanic rock deposits, both have neutral pH, and are both lightweight yet durable. Perlite tends to be sold in smaller granule size than pumice, so you could consider using large pumice granules as a base layer, and perlite as a standard mix component. Of the two, I marginally prefer pumice because it is slightly denser, more absorbent, and turns a shade darker when wet.

A disadvantage with both pumice and perlite is that they are less dense than most other soil components. As such, after many months or years of watering, the white granules can gradually become displaced to the soil surface. It is then less easy to detect when your bonsai needs watering, and with constant humidity, those granules accumulate algae on their surface. Your white top soil gradually turns green. This is not an issue horticulturally, but some people don't like its appearance.

Vermiculite

Vermiculite has a neutral pH. It absorbs up to 16-times its weight in water. It also provides aeration for the first few months; it is therefore very good for short-term development of fine roots. I tested new seedlings in pure vermiculite compared to various other mixes, and in one year, the seedlings in vermiculite excelled in root growth and overall vigour. However, during that year, the vermiculite had mostly decomposed into an almost homogenous mushy substance.

For this reason, vermiculite is not recommended for bonsai; it breaks down too quickly. You would have to repot every tree, every year. I use vermiculite only for new seedlings which I know will be repotted within one year. If you do decide to use vermiculite, mix it with a sturdier component, and plan to repot after only one year.

Soil mixes for different climates

If you live in a **hot, arid region** where water evaporates rapidly, use a **higher proportion of water-retentive components** in your soil mix, like akadama and chopped pine bark.

Warm-climate soil mix:
50% Akadama, 25% Pine bark, 25% Kiryuzuna

Conversely, if you live in a **colder, humid region**, your soil should have a **higher proportion of sturdier components** like kiryuzuna, pumice and lava rock, to prioritize drainage and structural integrity, especially through repeated freeze-thaw cycles. Don't use gravel, unless there are no other sturdy components available.

Soil layers and stratification

In **warmer climates**, it is **beneficial to stratify the soil into layers**, with larger granules forming the base layer (6-8mm or 1/4"-1/3"), then a mid-layer of standard soil size (4-5mm or 1/6"-1/5"), and a top dressing of small granules (2-3mm or 1/12"-1/8").

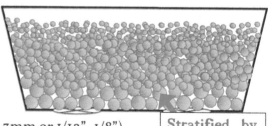

Stratified by granule size

Soil wicks water upwards against gravity, due to capillary action. Smaller particles have more wicking and longer retention than larger particles. Therefore, stratifying upwards from larger to smaller particles helps to maintain a more even water distribution vertically within the soil.

The base layer of larger granules is sometimes called an "aeration layer", or a "drainage layer". However, neither term is strictly correct, because this layer raises the "**perched water table**" – a saturated layer in the soil – which, in fact, limits both drainage and aeration. In hot, dry climates, a raised water table in the soil can be beneficial for short periods of an hour or two, but not in cool, wet climates. See *Horticultural aspects of pot size*, page 119.

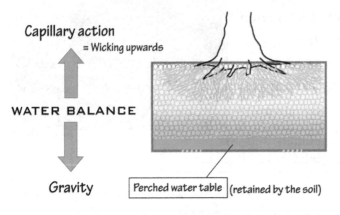

There is a definite advantage in using harder granules like pumice for a base layer: they prevent smaller particles from clogging the drainage.

So, in **wetter climates, don't stratify the soil in layers**, as drainage is more important than retention. Raising the perched water table is not good when the soil is constantly damp due to natural conditions. Hence, uniform particle size is preferable in cooler, wetter regions.

Summary – bonsai soil

It's most beneficial to combine sturdy components like pumice with absorbent soil types such as akadama or pine bark, in proportions that are appropriate to your climate. Such a mix ensures both retention and structural durability in your soil. Remember:

The ideal bonsai soil retains water and nutrients, while allowing adequate drainage and maintaining its structural stability.

Styles

See *Bonsai styles and styling*, page 50.

Tools

See *Bonsai tools*, page 62.

Tourniquet

A tourniquet is a length of wire that we tie around a section of a tree, such as the lower trunk, or just beneath an air layer. We do this in order to restrict growth beneath a certain trunk height, in order produce a new set of roots just above that height. Using a tourniquet is a safe way to do this without risking the life of the bonsai.

Don't tie it on so tightly that it cuts right into the cambium, restricting the flow of sap in the phloem tubes, or worse, the inner xylem tubes. If we restrict the sap flow from day one, we might as well go "all out" and ring-bark the tree – making a ground layer – an air layer near ground level, which is often successful, but risks killing the tree.

Twisting the tourniquet wire ends

So, the tourniquet allows sap to flow, but is *just* tight enough to prevent the trunk from growing thicker at that point.

As the tree grows over the following two or three years, the trunk starts to bulge above and below the tourniquet. The wire gradually cuts into the bark and cambium layer as the trunk naturally thickens. Provided the tourniquet is buried in soil, the tree's natural stress response is to send out new roots just above the tourniquet. This produces a ring of radial roots around the perimeter, which can develop into attractive surface roots over time. See *Nebari—the surface roots*, page 115.

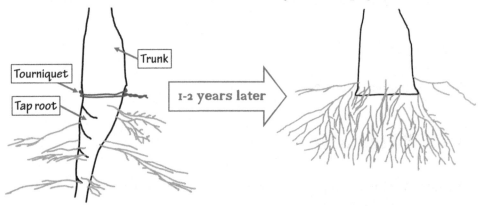

After one to two years, when enough fine roots have grown from above the tourniquet, you can safely chop off the tap root and remove the tourniquet.

The tourniquet also causes a bit of natural trunk flare, as part of the trunk pushes outwards above the wire. Trunk flare is a desired feature in bonsai, for various reasons. See *Trunk flare and root flare*, page 116.

Tourniquet failures and how to avoid them

Callusing

As the trunk thickens above and below the tourniquet, the bark starts to callus, and this can grow over the wire, completely engulfing the tourniquet and leaving an ugly bulge at that level, without roots.

 Use a calibre of wire thick enough to **prevent the tree from callusing over the wire**, for example, 3mm wire (1/8").

 Wrap the wire double or triple around the trunk, to provide sufficient distance between the lower and upper bark, so that as it calluses, it can't completely cover the wire.

Twist the wire ends together with a sufficient number of twists to ensure that the trunk's outward pressure on the wire doesn't cause it to untwist and become loose.

Insufficient roots above the tourniquet

If the tourniquet is allowed to stay at soil surface level, the tree won't send out roots above the wire. The bulge below the wire becomes exaggerated and sends out more roots beneath.

 Bury the tourniquet **at least 2.5 cm (1 inch)** beneath the soil surface, or significantly deeper if it's a thick trunk.

Uneven roots above the tourniquet

If you wrap the tourniquet beneath existing roots, these older roots may continue to grow thicker much faster than any new roots that emerge. After a couple of years, this imbalance will become increasingly exaggerated as the older roots thicken and supply the tree with a lot more sap than the newer roots.

 Remove any roots above the level of the tourniquet, provided there is sufficient root mass below it.

The old roots must all be below the tourniquet, so that only new roots grow above it. The new root spread will grow more evenly this way.

Exception: If your tree has previously sent out a spread of adventitious roots from above the soil line, you might decide to convert these into the new main root system, as explained in *Ten reasons to repot your bonsai, reason 10* on page 134. To do this, wrap the tourniquet around the trunk beneath the adventitious roots, to gradually cut off the supply from the old roots, then build up some deeper soil around it for a few years.

Trunk chops, trunk thickness and taper

Performing a major trunk chop is a crucial part of bonsai development; otherwise the tree would grow upwards for many decades or centuries. Before discussing trunk chops, let's look at the trunk girth or thickness.

Trunk thickness

A trunk's thickness is measured as its width just above the root collar.

To thicken the trunk of a tree or shrub as quickly as possible, plant it in the **ground, or in a large, deep pot**, and let it grow unhindered for several years. Young trees rapidly maximize energy production by pushing up long apical shoots to get as much foliage area in sunlight as possible.

Deep pot or in the ground? For the first five years or so, growth rates are similar. However, pots eventually fill with roots, so the growth slows compared to ground-planted trees. Also, trees in the ground are less vulnerable to wind, ice and drought. Despite this, I still use large pots because they are convenient, moveable, and repotting is simpler.

Grow it fast or slow? Here are some disadvantages with growing trees either in the ground or in large pots to thicken their trunks quickly:

- The inconvenience of growing a massive tree for a small bonsai.
- Unwieldy growth can temporarily ruin a refined bonsai. Rapidly grown apical shoots overshadow lower branches, repressing the compact interior growth that we want on a small bonsai.
- The root system grows thick, deep woody roots. Eventually, these must be "tamed" and trained to fit in a small, shallow pot.
- While the trunk thickens quickly, it grows with almost no taper.
- A fast-grown trunk can eventually develop chunks of cracked bark that appear disproportionately big for a small bonsai, whereas trees grown in small pots (with frequent pruning and repotting) gradually form tiny cracks in realistic proportion to the overall tree.

With all those disadvantages, it may be more practical to grow it slowly in a small pot. Or buy a pre-bonsai that has already been thickened.

Chopping the trunk

If the trunk is not too thick, use **sharp garden shears** for a straight, clean cut. If the trunk is thicker than about 1.5cm (3/5"), use a **branch saw**.

But how do you decide when to chop the trunk? And how low to chop?

Well, chopping the trunk not only cuts the height; it also stunts the thickening rate of the remaining trunk, because less photosynthates are generated by less foliage. So the "right time" to chop is when **you judge the lower trunk to be nearly thick enough** for the bonsai you want. But how low to chop? See *Trunk chop succession over many years*, page 175.

Deciduous and broadleaf evergreens

For deciduous trees, and some broadleaf evergreens, chopping a tree's trunk is just a severe form of pruning. Provided the tree is healthy and vigorous, it can take a big chop and will bounce back with new growth.

You can do it in late spring after the first flush of leaves has hardened, or early winter when the tree is dormant with no sap flow, to avoid "bleeding". Be aware that in winter, the chop wound takes a few more months to heal because the tree stays dormant until spring.

Trunk chop with a branch saw

In winter, dab on some cut paste, especially around the perimeter of the cut where the circle of cambium has become exposed to the elements. The tree's own sap provides a good wound sealant; however, in winter with mist and rain, there is a slight chance that harmful pathogens, such as bacteria or fungal spores from other plants, could enter the cambium and grow in the tree's vascular system. See *Cut paste*, page 76.

Coniferous trees

> **Caution:** Chopping the trunk too low on a conifer can kill the tree. Be sure to keep **at least one third** of the coniferous foliage alive.

Trees such as pines, junipers and spruces, can be trunk chopped in late autumn after the tree goes dormant, or in early summer, when the energy level is high and back budding can occur. In summer as it bleeds resin, this can provide a good natural wound sealant, while in winter you should use a commercially available cut paste or putty for conifers.

Rather than directly chopping off a conifer's trunk at the desired height, it is sometimes more artistically pleasing to strip some branches and leave a section of dead trunk on the tree. You then have the option to remove the bark, and carve a deadwood feature on the trunk sometimes called "*shari*", which can look fantastic on old conifers. The contrary, completely removing it now, denies you that option later on.

Deadwood on a juniper

Trunk taper

Trunk taper is one aspect that helps a bonsai appear like a mature tree.

More trunk taper makes a bonsai appear to be viewed from nearby, as if looking up at a tree. **Less taper** appears as if viewed from a distance.

To increase the overall trunk taper:

- Grow large sacrifice branches low down on the trunk.
- Prune the apex and upper branches harder and more frequently than lower branches.
- Develop a radial root system, to thicken the root collar and grow trunk flare. See *Nebari—the surface roots*, page 115.
- Use a succession of trunk chops over multiple years. Read on…

Trunk chop succession over many years

Here is a succession of trunk chops that could take up to 15 years, or longer, to appear natural. The first chop or two, the tree could still be in a large development pot, to avoid slowing down growth in a small pot.

First trunk chop: Chop the trunk leaving a small branch or shoot to grow as the next leader.

One year later, carve the chop diagonally, sloping towards the rear view.

3 to 5 years later Chop the trunk just above the next branch up

Another 3 to 5 years later   **A further 3 to 5 years later**

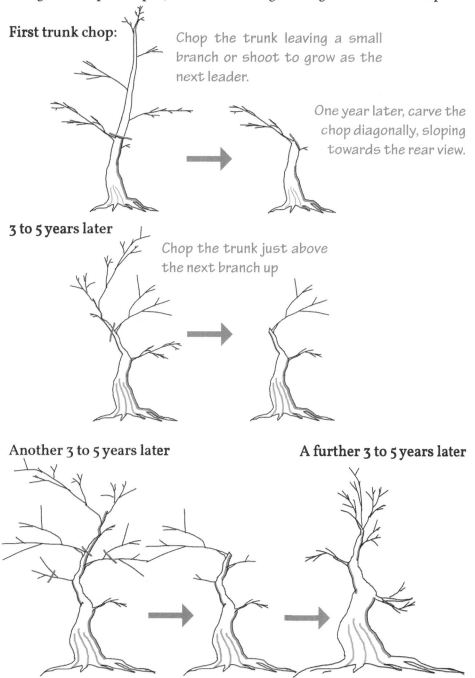

Watering bonsai

Water is, of course, the elixir of all life. And it is the one thing that no tree, anywhere, can survive without.

As long as you have free drainage through the whole pot and beneath, and soil particles of sufficient size to ensure always fresh oxygenated water is percolating down easily through all of the soil, you can water a bonsai to your heart's content. In these healthy conditions, there is no such thing as "overwatering" because all the excess water drains away.

By contrast, when we don't water enough – **underwatering – is by far the most common cause of bonsai deaths.** When there is not enough water available to the roots, the tree's stress response is to close the stomata under the leaves to stop transpiration, reducing the amount of water evaporated. If the situation continues, some or all of the leaves are jettisoned to conserve the limited remaining water in the trunk and roots. See *Reviving a dying bonsai*, page 154.

"Overwatering" *can* damage a tree's health if sodden soil is in contact with the roots and loses its ability to drain, starts to stagnate, and grows harmful bacteria. When this happens, you can smell the bad odour underneath a pot. When the harmful bacteria start to attack the roots, the tree starts to die slowly, at first with no new growth and a decline in leaf vigour, followed by leaf loss, and eventually tree death as the bacteria spreads through the root system.

Therefore, **always ensure your pot is draining through** when you water your bonsai, and don't leave it standing in a puddle of water.

Always ensure good drainage

How much water does my bonsai need?

The amount of water needed by a tree depends on its size, the species, the time of year, the air temperature and humidity, the total leaf area, and the water-retentiveness of the soil. As you can see, it would be impossible to give one watering guideline for all bonsai.

In **hot summer temperatures**, most of the water sucked up through the roots arriving at the leaves, evaporates off through the stomata to keep the tree cool. In these conditions, **much more watering** is needed to keep the tree healthy, and not much of that water goes to new growth.

So, in **summer**, as long as you have good drainage in your pot and soil, **water your bonsai more than once a day**. On the hottest days, in a dry, hot desert-like climate similar to Madrid, water **three times a day**.

Water a lot more in summer

In **spring** and **autumn**, with lower temperatures, trees don't need as much water, because much less is transpired through the leaves.

In the cold **winter**, your bonsai can go for days without watering. If overnight freezing is expected, don't water late in the day. A water-filled pot could crack overnight due to water-ice expansion. Watch out for a prolonged freeze because roots can't absorb water from ice, so some trees could die of underwatering. See *Overwintering*, page 29.

In general, **let the top soil be your guide:**

When the top soil starts to dry, water your bonsai.

This is why, when I repot my bonsai, I use a **top dressing** for the soil that **significantly changes colour tone between wet and dry**. This way I can easily see at a glance when a bonsai needs watering.

In shallow pots, this is more important than ever! If you leave a shallow pot dry for long, you'll lose your tree. See *Horticultural aspects of pot size*, page 119.

Whatever your climate, when you water, **ensure that all the substrate gets thoroughly wet** without leaving dry pockets of soil.

 Water all of the soil surface until a substantial amount drains through the pot and out of the drainage holes. For trees with a wide trunk base, water the entire trunk base sufficiently to ensure the soil underneath is wet.

Drip trays

Indoors, you might need to water your bonsai on a drip tray. If you do use a drip tray, don't let the pot stand in a puddle of water—you could use the excess to water other houseplants.

Alternatively, place a layer of gravel on the drip tray, so that there is constant humidity, without the pot base being permanently submerged in the puddle. For this reason, some people call the drip tray a "humidity tray".

Drip tray

Don't use a drip tray as a method of "underneath watering". The water might not reach the upper layers of soil, while the base layer becomes slowly stagnant in a permanent puddle of water.

Watering guidelines – summary

The exact guidance for watering depends on:

- the **bonsai size**. Smaller trees in tiny pots dry out quicker than larger trees in big pots.
- the **species**. Some tolerate drought more than others.
- the water-retentiveness of **your soil**. In a hot, dry climate use soil components that retain more water, like akadama and pine bark. In a cold and wet climate use more draining materials like pumice and lava rock. See *Soil*, page 161.
- your **climate,** local **microclimate** and current **weather** conditions.

However, as a very general rule for most bonsai you can follow these guidelines. Certainly, in summer, always err on the side of more water, provided you're using granular soil with good drainage.

Summary:

If you live in a	Warm, dry climate	Colder, wet climate	Tropical climate
Spring	Water daily	Water when the top soil gets dry	Water when the top soil gets dry
Summer	Water 2 – 3 times a day	Water daily	
Autumn	Water daily	Water when the top soil gets dry	
Winter	Water when the top soil gets dry	Water if the soil is dry. Don't water evenings.	

Watering systems compared

Here are five methods of watering, with pros and cons of each. I use all of these, except the third method, dunking.

Question: What can keep a bonsai alive forever? **Answer...**

1. Watering can

The "good old" watering can is probably the simplest and most commonly used method of watering bonsai.

For a watering can to be appropriate for bonsai, it should have a rose with fine nozzles, that releases many fine streams, rather than few thick jets of water. A heavy torrent of water can significantly disturb your bonsai soil, potentially unearthing roots, or burying your fine top soil underneath the thicker soil mix.

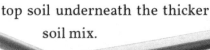

Advantages: You can carry a watering can to any remote location, perhaps out of reach of a hose pipe. It's easy to add liquid fertilizer supplements into the water. It is also easy to use collected rainwater, by dipping the body of the watering can into the bucket of water. And, it's the most convenient method of watering indoors.

Disadvantages: If you have many trees to water, you might need to refill the can several times for each watering. Also, the fine spray nozzles can get blocked by particles in the rainwater. To unblock them, soak the whole rose in hot water and vinegar, then use a very fine pin to dislodge the residue from the insides of the holes.

2. Watering hose

Like the watering can, a bonsai hose should be fitted with a bonsai-appropriate spray head, with many fine nozzles, to avoid the water moving the soil granules around in the pot.

Hose with fine-spray nozzle

Advantages: It's a quick job to water a lot of bonsai trees with a hose.

Disadvantages: Adding liquid fertilizer is impractical, or at least very inconvenient through a hose. Also, you are limited to the distance of the hose length from your tap. And it's not a practical solution for indoors; watering with a hose makes a lot of splashes and run-off water.

3. Dunking the pot

Fill a sink with water and submerge the entire pot and soil under the water for more than 5 minutes, to ensure that all the soil gets fully wet.

Advantage: Some professionals recommend this method to mitigate the possibility that watering with a can might not thoroughly wet all of the soil, for example underneath a thick trunk base.

Disadvantages: The dunking method is very inconvenient if you have more than a couple of bonsai to water, and it is inappropriate for very tall pots. The water can rearrange your soil mix, shifting the lighter components such as pumice and perlite to the surface, while pushing the smaller and denser granules downwards. Another side effect is that minerals and fertilizer supplements in the soil can partially dissolve and sometimes get deposited on the pot lip and top soil.

4. Water box

This method consists of a highly absorbent felt strip passed through the drainage holes of the pot, and draped into a plastic box full of water. One commercial name for this contraption is "hydrobonsai".

Placing the felt strip into the water box

Advantage: Good for a shohin-sized indoor bonsai, for example, while you are on summer vacation. Keeps the soil damp for up to two weeks.

 Position the bonsai and water box away from direct sunlight, to avoid heating the water and potentially cooking the roots.

Note: This method should not be used for long-term watering needs. It is also not recommended for larger bonsai that consume more water.

Disadvantages: In the long term, the roots grow into the felt and start to block the drainage holes. With the drainage compromised, and permanent moisture from the felt strip, the soil around the base of the pot gets stagnant, and the roots in that stagnant soil start to rot. You may notice a putrid smell coming from the underside of the pot, or the tree starting to shed leaves. See *Reviving a dying bonsai*, page 154.

5. Automatic watering system

If you sometimes need to go away for a few days, and don't want to rely on someone else to water your trees every day – in summer twice a day or more – I recommend installing an automatic watering system.

Advantages: Even if you're not planning to go away, this can save you a lot of effort in the long term. Also, if for any reason you can't get to your benches on a given hot day, you know that your trees will be watered.

Disadvantages: If you live in a humid or rainy country, leaving your watering to an automatic timer could be pouring a lot more water onto your soil than is needed. This can be wasteful in terms of water consumption.

Using excessive water should not harm your tree as long as you have good drainage through the soil and under the pot. However, if you notice the water is taking a long time to drain through, it could highlight an issue in the soil or the pot, and you should consider repotting or slip potting. See *Repotting: why and when*, page 129.

Planning an automatic watering system

1. Check the water pressure

Firstly, check the water pressure at the tap or faucet that will serve your watering system. The ideal pressure is 1.5 to 2 bar (about 20 to 30 psi).

If it's much lower, your watering nozzles might not get enough pressure to function properly. If it's much higher, you risk bursting a pipe or rupturing a joint in your system. This would have dreadful consequences, not just because it spurts gallons of wasted water out of the break, but worse, it leaves your bonsai pots to get dry.

Remember that water pressure is not the same as flow rate; if the pressure is too high, even a trickle can blow a joint. If the pressure is 2.5 bar (40 psi) or higher, install a pressure reducer between the tap and your hose pipe. Pressure reducers are cheap and could save your trees.

If you can't get hold of a pressure gauge to measure your water pressure, there is an alternative, very approximate method. Hold a hose pipe pointing upwards, with the tap on. Hold your thumb over the end of the hose, until you feel the pressure build under your thumb. Let a tiny jet of water out and estimate how far upwards it reaches. If it's around 6 feet or a couple of metres high, your water pressure is probably okay. If it shoots up as high as a two-storey house, there is too much water pressure, and you'll need a reducer.

2. Choose a timer

Look at reviews online to choose the most reliable timer. It's worth spending extra money on this, to avoid heartache someday in the future.

Although wi-fi connected timers are available, I personally don't want to risk the unreliability of a wi-fi connection. If it disconnects while I'm away, will my trees still be watered? How will I know?

I recommend **digital programmable timers**, because you can set the exact times for watering, as opposed to an interval timer which simply waits a set time before watering. For example, a programmable timer can be set to water at 10am and 3pm. But in this example, watering on an interval would be 10am and 10pm, so in summer your trees would go without water all afternoon and evening.

3. Group your trees into watering zones

If you have more than about 30 trees, it is worthwhile dividing them into groups, and setting up a separate watering zone for each group. The main purpose of this is to ensure that there is enough water pressure for all the nozzles to function properly for each group of trees.

4. Choose your watering pipes

Programmable timers for three watering zones

Use wider pipes to transport the water over longer distances; thinner tubes to connect from the thick pipe to the bonsai pot. Thin tubes resist the water flow significantly more than wider tubes, which is acceptable for a short distance to one or two pots, but would seriously impede the water flow over longer distances with many nozzles.

5. Choose your nozzles

Small pots: Use adjustable drippers. Use a minimum of two drippers per pot, so that if one of them becomes blocked the other still provides water. You can use up to eight drippers per thin tube.

Adjustable drippers

Mid-to-large bonsai pots: Use spray nozzles. Again, use a minimum of two nozzles per pot. Spray nozzles produce a fine mist, with a lot more water than misting nozzles (which are for ambient humidity rather than for irrigation). Spray nozzles evenly cover a good area of soil, which ensures all the soil is watered without disturbing the granules. Use a maximum of four nozzles per thin tube.

Spray nozzle

Large, deep grow pots: Use adjustable sprinklers secured in the soil with their own plastic stakes. Each sprinkler can produce up to eight times the water flow that an adjustable dripper produces. This means that one watering zone may contain far fewer sprinklers than drippers and spray nozzles. For this reason, it's best to have all the sprinklers in their own zone. Use only one sprinkler per thin tube.

Sprinkler

Small grow pots: Group together many small grow pots, and suspend a few spray nozzles over a larger area of pots. I use between three and four nozzles for each group of eight pots. I also use square pots so that their edges touch, and water isn't wasted in gaps between the pots.

Suspended line of spray nozzles

Installing the watering system

If you have a lot of trees, this is a big project. But it's time well invested in the long run. To install a new irrigation kit, estimate a **minimum** of **fifteen minutes initial setup time per bonsai**. Probably longer. Add at least a couple of hours for securely installing the tubing.

1. First install the pressure reducer, main hose, and timer.

2. Attach the wider tubes or pipes to the outlets of your timer. If a pipe goes a long distance, use brackets or ground pins to hold sections of the pipe in place.

3. If you have a bonsai bench, use brackets to secure the pipe to the rear side of the bench. Use corner joints to turn 90°; don't risk bending the pipe, because if it kinks, it can restrict or stop the water flow at the crimped point.

4. Use hose clamps to secure sections of pipe to the joints. Insert an end plug, and clamp that too, so that it doesn't blow out.

5. Test the timer and water in the pipe, to ensure that there are no leaks before attaching the thin tubes to deliver water to the pots.

6. For each bonsai, insert a short thin tube into the wide pipe. Most kits include insertion joints that can pierce the wide pipe, staying securely inserted. However, it greatly helps to first pierce the wide pipe with a sharp point at the position you need to insert the tube.

One wide pipe, multiple thin tubes

Installation tips

- Use a longer section of narrow tubing than you think you need, exceeding the distance between the pipe and the bonsai. If the tube is too long there's no problem, but if it's too short then it's of no use.
- Use a cup of boiling water or a lighter flame to briefly heat the tube end before inserting an attachment. Heating the tube makes it more flexible, and slightly expands the hole. Then, as it cools, it shrinks onto the inserted plastic, securing it firmly onto the joint.
- For spray nozzles, position the tubes between the bonsai pots, so that you can direct nozzles at the soil from both sides of each pot.

- Attach metal supports to the bench to hold the tubes over the pots. Use bonsai wires to position each tube and to direct the nozzles.
- If your timer has already been operational for a few months or more, **change the batteries in the timer before you travel**, even if you think the batteries still have charge. Do this well in advance of your trip so that you have time to check that the new batteries work.

Support holding tubes in place

- After adjusting your outdoor timers, cover them with a large waterproof sheet like an upside-down plant tray, to shield them from direct sunlight and rain—even if they are waterproof.

Testing the system

At least two weeks before you plan to travel, put the system into daily operation. Set it to water exactly as you want it to operate while you're away. You'll probably need to adjust some nozzles or the timings to optimize your system. Don't wait until the day before you travel because that's too late to verify that an adjustment has worked as you intended.

Optional protective measures

Use these measures for additional peace of mind while you're away:

- **Security camera:** Install a wi-fi-connected security camera, and point it at one of your favourite bonsai, so that you can check it at scheduled watering times.
- **Second system:** Install a second watering system completely independent of your first system, including a different main tap, timer and pipes. Although this is a "belt and braces" security measure, it protects your trees in case one of the systems fails.
- **Trusted person:** Ask someone who you trust to visit daily at one of the scheduled watering times, to visually check that all the pots are receiving water. This is far less of a responsibility than it would be to water all the trees two or three times a day in summer.

Wiring branches

The quickest way to reposition branches permanently is by using wires for a period of time. When the xylem pathways in the branch set in the new position, they lignify, creating a woody interior to the branch that stays in that position. This could take a few weeks, or many months, depending on the species, the branch, and the time of year.

Wiring can be done at any time of year, but keep in mind that branches only start to lignify when they are actively growing. For example, it is simpler to wire a deciduous tree in winter without leaves; however, its branches won't start to set in the new position until sometime in spring.

Wiring a branch does not damage it, as long as the branch doesn't snap. For this reason, apply the wires first, then start bending the branch with your hands. Don't bend while applying the wire, because you can't feel when the branch is about to break. The wire's job is to hold the branch firmly in the new position; not to move it there.

Helical (spiral) wiring

Wherever possible, wire branches in pairs. This is both efficient, and works well with bifurcating branches. Most of all, it creates a firm pivot point between the two branches in order to lever them in the desired directions. If there isn't a nearby second branch, use the parent stem – such as the trunk – as your anchor or pivot point.

Using this Hornbeam's trunk as the pivot

What thickness of wire should I use?

There isn't a simple rule for the gauge of wire to use: it depends on the species, how flexible the branches are, and the wire material. Use thicker wires to bend thicker stems, and finer gauge wires to bend thinner stems. As a very approximate guide, choose a wire that is about half the thickness of the branch section that you're bending. For very strong, stiff branches, it is sometimes necessary to double wrap the wire, creating a double helix on the branch.

How to wire a bifurcating branch

Work on pairs of branches. In a bifurcating branch system that gets thinner and finer, start by wiring the **outermost pair of thickest stems** first, using an appropriate thickness of wire. Afterwards move on to successively finer pairs of stems, with correspondingly thinner wires. This method is the neatest and most effective way to wire the entire branch, using appropriate gauge wires for each section of the tree.

1. Use a length of wire **significantly longer than twice the branch span** you're wiring, to allow for multiple wraps around it.

2. Loop the middle of the wire either over or under the parent stem as follows:

    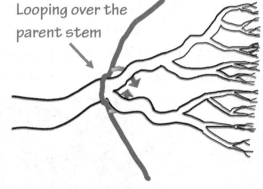

    - **To bend the branches upwards, loop the wire over** the parent stem (see diagram, right).

    - OR, **to bend the branches downwards,** loop the wire **under** the parent stem.

    This loop on the parent stem serves as the pivot point, for the wire to apply opposing force to the branch.

3. **Use two hands**. With one hand, firmly hold the wire that's already in place, while with the other hand twist the loose wire around the branch, making a continuous **45° angle** against the direction of the branch.

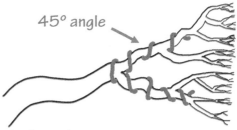

Ensure the wire **lightly touches the branch** without pressing into it.

As you apply the wire, twist your hand in the direction of the spiral for a smooth, continuous flow. Move your other hand up after each twist to firmly grip the wire that you just placed. Adjust the angle if needed to avoid wiring over lateral buds or leaves.

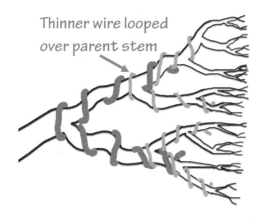

4. For thinner branches, use correspondingly thinner wires. Apply the same technique as before, further down the branch: loop the wire around the parent stem on the same side (over or under) as the first wire.

5. **Always wire the outermost branches** in each successive branch pair. This ensures a smooth spiralling flow from section to section without crossing wires, and also simplifies wire removal. Repeat this process down to the smallest branches or twigs.

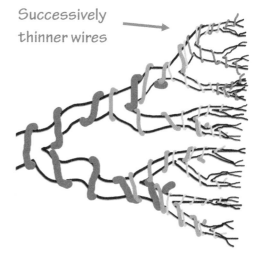

6. If a wire extends past the end of a twig, you can either snip the wire back to size, or bend it back to create a loop at the end. These loops give you the flexibility to uncoil the wire ends after a few weeks and use them for further wiring as the branch tips grow longer.

Wire bite

As stems grow thicker quickly during the growing season, the wire doesn't magically loosen on the branch, so the bark starts to push outwards at the wire. After a time, the wire appears to bite into the expanding bark. Avoid this if possible, because the spiral-shaped callus scars in the bark look artificial, and can take up to ten years to grow over naturally—if ever.

Ugly wiring scars

 After wiring, **check the wired branches at least once a week** to ensure the bark isn't starting to press into your wire. Remove the wire at the first sign of branch thickening.

Protecting the stem before wiring

If the stem is thicker than about 8mm (1/3"), before wiring, tightly wrap it with strips of wet raffia, or cotton twill tape. These protect the bark from wire bite, and allow more flexible bending before the stem snaps.

Note: Covering the stem will prevent back budding in that area.

Twill tape is easier to apply, but is more conspicuous on your bonsai. You can stain the tape or raffia with black tea to make it less noticeable.

Guy wiring

Branches sometimes need to be brought down to horizontal, or pointed diagonally downwards. This simple change can make a small bonsai look more like a mature tree, with massive, heavy branches gracefully succumbing to the pull of gravity.

Guy wires on a Ficus microcarpa

Guy wiring is an effective method of pulling branches downwards without creating the tell-tale spiral marks that helical wiring sometimes leaves. You can keep a guy wire in place for over a year, as long you use a flexible plastic sheath to protect the bark where it is pulling.

Tip: **Strengthen your guy wire by doubling it,** looping it around both end points, and tying the ends of the wire together. Then, use a thin rod as a key between the two parallel wires to twist them together, thus shortening the length and leaving a small gap in the middle. Using a **twisted pair lets you adjust the tension** and tighten the wire over time.

Anchor points

To secure the wire firmly and provide sufficient downward force, you have three options: use a thick surface root, tie it to another wire under the pot, or fasten a screw into the lower trunk as your anchor point—if the trunk is thick enough to support a screw and the tension.

If you **anchor a guy wire to a surface root**, choose a strong root that is a lot thicker than the branch so that the equal and opposite force does not gradually pull your root upwards over time.

If you **anchor the guy wires to the pot**, aim to wire down an equal number of branches on both sides of the pot, to apply the force evenly. Without this balance, over time the uneven pulling force could potentially tilt the entire tree and root system in the soil towards the side with more tension.

Guy wires balanced left and right

Cascading guy wires

You can also wire thinner upper branches to thicker lower branches, hence pulling down the upper branch further than the lower branch is pulled up. Ensure the lower branches are thicker than the upper ones. You can wire branches like this, cascading all the way down the tree, so that ultimately, only the lowest branches require tying down to the pot or trunk base.

Cascading guy wires (all twisted pairs)

Copper wire or aluminium wire?

Copper and aluminium are the two most used metals for wiring bonsai. Aluminium is cheaper than copper for the same wire thickness (gauge). However, copper wire is stronger than the same gauge of aluminium.

Therefore, it might sound more attractive to use copper, since you can achieve the same bending force with a thinner copper wire than aluminium. Indeed many seasoned bonsai artists use copper wire, but it comes at a price—apart from higher cost. When copper is bent, its internal structure dislocates, causing the copper to stiffen even more. This makes it more difficult to remove from branches than aluminium, so you have to snip it off. Even if you carefully twist it off, the hardened copper is more brittle, with kinks, so it is only good for one-time wiring.

For the above reasons, **I recommend aluminium for spiral wiring**. When it's time to remove it, in many cases you can uncoil it to use it again, and smooth out the kinks. However, if you find that removing the wire risks scraping the bark or damaging some twigs, snip the wire off with a fine tipped wire cutter. Reusing wire is good, but not at the expense of your branches.

Coated aluminium wire

For **guy wiring, I recommend copper**. Guy wires can be starkly visible in a bonsai planting as they span the space between branches and pot, so it's a good idea to use thin wires. Copper is strong under tension, and hardly stretches at all, so you can use surprisingly thin wires for guy wiring, less than a millimetre (1/25").

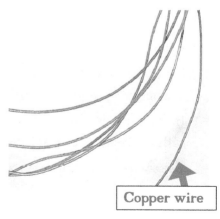
Copper wire

When to remove wires

Remove spiral wires **before** they begin to cause wire bite. You can always rewire if needed, so there is no penalty for removing them early.

Guy wires can be left on permanently if needed.

Wiring the trunk

In the early stages when a tree is young and thin, you can artificially produce movement in the trunk by applying aluminium or copper wires spiralling around it. This is a quick way to develop a bonsai in various styles, such as informal upright, twisted trunk, yamadori, cascade, or literati. See *Bonsai styles*, page 51.

Disadvantages of wiring the trunk

- Young tree trunks thicken rapidly, often producing the tell-tale **unsightly wiring scars** that can take many years, if ever, to conceal.
- A starkly curved trunk **may appear more like an artificially-shaped** bonsai than a realistic miniature tree. The alternative, a series of trunk chops, can look more natural but takes many more years to develop. See *Trunk chops, trunk thickness and trunk taper*, page 172.
- **Minimal trunk taper** compared to a series of trunk chops.

Tips for wiring the trunk

- Choose a young whip with a stem diameter thin enough to remain flexible, considering the species. For example, a brittle species like boxwood shouldn't be thicker than about 4mm (1/6"), but a flexible species like juniper or pine could be 8mm (1/3") or even thicker.
- If the trunk is thicker than about 8mm (1/3"), first wrap it with wet raffia or twill tape. See *Protecting the stem before wiring*, page 192.
- Use a wire gauge approximately half the thickness of the trunk.
- If you choose to keep the whip in its original pot, **poke one end of the wire deep into the soil** to the bottom.

- If you're repotting it, before adding soil to the new pot, **fix the wire to the pot's base**. This creates a firm anchor that enables you to bend the trunk right from its base.

Wire fixed to pot

> **Note:** Apply the wire first; bend the trunk after. Don't attempt to bend the trunk while applying the wire.

- Firmly hold the wire at the base of the trunk. Use your other hand to bend the free end.
- Spiral the wire around the trunk at a **continuous 45° angle**.
- Be careful **not to tear roots off the trunk** while doing the first bend.
- **Shape the trunk** with a series of curves, or a continuous curve **in 3 dimensions**, with left-to-right and front-to-back movement.

Bending the wired trunk

- For coiled "yamadori" style using a conifer like pine or juniper, on **the first day bend it only moderately** towards the desired shape. Then a week later, bend it more forcefully into shape. These species become slightly more supple a few days after the initial bend.
- Be prepared to apply a second wire along the same spiral, in case a single wire is not sufficient to hold the bends. Don't poke the second wire right through the soil; it can start near the trunk base.

Double-wired trunk

- **Check the wiring weekly**, and remove it at the first sign of trunk girth expansion. Reapply the wire if necessary.

Advanced techniques

In my tutorials, I have also covered many of the following subjects, which are not in the scope of this volume of the guide:

Carving trunk chops
- Creating more taper – carve a trunk chop into a V-shape to give the illusion of more taper on one side and a branch on the other.
- Hollowed trunk – to appear as an old tree hollowed by nature.
- Van Meer technique – to fold the cambium and bark over a chop.

Dead wood
- Jin – dead branch carving
- Shari – trunk carving

Extreme branch bending
- Branch splitting – using a branch splitter at the bend site
- Wedges – using a straight branch cutter to cut wedges from the inside of the bend

Fusion
- Tying together several young whips of the same species so that their trunks gradually fuse and start to grow as one tree

Grafting branches and roots
- Approach graft
- Thread graft
- Root graft

Ground layer – ring-barking for a better trunk base and future nebari

"Yamadori" – collecting trees from the wild

Index

acorn: *See* propagating
air layer: 38, 126-28
akadama: 163-65, 167-68
   in an air layer: 128
apex
   creating: 80-81
   pruning: 68-69, 78
apical dominance: 36-37, 69, 87
automatic watering system: 183-88
   on a bonsai bench: 44
auxins: 36-38
   and back budding: 66-67
   and root growth: 124, 126, 131
azalea: 37, 69, 78, 103-4, 163
back budding
   encouraging: 66-67
   repression: 36, 68, 192
beginners' quick guide: 15-16
bench: 40-44
berries: 106
bifurcation: 67-68, 78, 189-92
bonsai soil: *See* soil
boxwood: 37, 59, 102-3
broadleaf evergreen: 28-29
   fertilizing: 100
   pruning: 70, 72
   repotting: 145, 150
   trunk chops: 173
broom style
   pruning: 69, 79
   styling: 51-52, 108
calendar
   branch pruning: 70-73
   fertilization: 100-101
   flowering and fruiting: 107
   repotting: 136-38
callus tissue: 34, 71, 74
callusing: 76, 170, 192
cambium: 34
   and plant recovery: 154-55
   in air layers and cuttings: 125, 127
   pathogens & cut paste: 74, 76, 173
carbon dioxide: 33, 85, 89
cascade: 17, 58, 107, 118
chumono (two-person tree): 45
clip and grow: 76-77, 79
colander: 121
colour
   flowers: 103
   leaves: 32, 155
   pot: *See* pot, choosing
   soil: 161, 164
conifers
   climate and location: 27-29
   cut paste: 76
   fertilizing: 100
   formal-upright style: 54
   pruning: 10, 75
   root pruning: 146, 148, 153
   trunk chops: 174
crab apple: 19, 105
cut paste: 76, 173, 174
cuttings: 10, 29
   taking: 124-25
cytokinin: 36-37
   and back budding: 67
   and root growth: 131
deadwood
   carving: 174
   cutting through: 62
   styling: 60-61
deciduous
   broom style: 52
   climate and location: 27-29
   fertilizing: 100
   pruning: 71, 73
   root pruning: 153
   trunk chops: 173
defoliation: 81-82
developing bonsai: 83-84

directional pruning: 77-78, 79
drip tray: 178
elm: 77, 124, 145, 152
    Chinese: 16, 30, 152
energy: 31-32
    balance by pruning: 68-69, 147-52
    calendar: 27, 70-73, 136-38, 150
    flowers and fruit: 102-4, 106, 157
    production: *See* photosynthesis
    reduction by defoliating: 81-82
evergreen
    broadleaf: *See* broadleaf
        evergreen
    narrowleaf: *See* conifers
fertilizer
    recommendations: 98-101
    science of: 85-97
ficus
    macrophylla: 115
    microcarpa: 16, 152, 193
flowers: 102-7
flush cut: 64, 74-75
foliage: *See* leaves
forest: 56-57
formal upright: 54
front of tree: 108-9
fruit: 102-7
fungus
    infection: 139, 156
    mycorrhizae: 146-47
garden: 40-44
gibberellin: 37, 88
grove: 56
growth regulator: *See* hormone
hawthorn: 106
hedge pruning: *See* profile trimming
hormone
    plant: 36-38
    rooting: 38, 125-27
indoor bonsai: 25-30
informal upright
    styling: 52-53, 107
    wiring: 196
insects

    and fertilizer: 90, 93
    and flowers: 102, 105
    infestation: 139, 155-56
internode length: 23-24, 65, 88
irrigation: *See* automatic watering system
jin: 198
juniper
    images: 17, 19, 56, 58, 174
    needle and scale foliage: 149
    pruning: 149
    repotting: 146
    wiring: 196
kanuma: 163-64
katade-mochi (two-arm tree): 45, 49
kiryuzuna: 163-65, 167
komono (one-arm tree): 45-46, 48
landscape: 60
lava rock: 163-64, 167
leaf
    biology: 32
    size: 23-24, 65
leaves
    defoliation: 81-82
    dropping: 148, 154, 176
    on cuttings: 124
    yellowing: 71, 90, 97, 154
leverage: 62
mame (palmtop bonsai): 45-47, 162
maple
    Japanese: 11, 23, 54, 124, 139
    trident: 20, 22
moss: 110-14
    in landscapes: 60
    propagating: 114
multi-trunk: 55
mycorrhizae: 146-47
nebari: *See* surface roots
nitrogen: 87
NPK: 86, 96
oak: 21, 146, 150
olive: 30, 124, 152
    wild: 52, 89
outdoor bonsai: 25-30

pathogens
   avoiding: 64, 74-76, 173
   eradicating: 147, 155-56
peat moss: 161-62, 164
   for cuttings: 125
penjing: 60, 112
perched water table
   and moss: 112
   and pot size: 119-20
   and soil: 168
perlite: 164, 166
pests: 155-56
phloem: 33-34
   and plant recovery: 155, 158
   in air layers and cuttings: 125, 127
   sap transport in spring: 137
phosphorous: 87
photosynthate: 32, 85, 124, 173
photosynthesis: 32-33
   and defoliation: 81-82
   and recovery: 157, 158
pine
   bark: 164, 167, 168
   pruning: 78
   repotting: 146
   wiring: 196-97
pomegranate: 106-7
pond basket: 121
portulacaria afra: 18, 47, 152
pot
   ceramic: 119, 121
   choosing: 118-21
   depth/width/size: 118, 119-20
   drainage: 10, 176, 178
   forest: 57
   repotting: 132, 135, 143
   training: 120-21
potassium: 88
profile trimming: 51, 57, 79
   roots: 143
propagating: 123-28
   by air layering: 126-28
   by cuttings: 124-25
   by seed: 123

pruning
   and wiring: 79
   apex: 68-69, 78
   branches: 65-79
   roots: 129-43
   tools: 62, 64
   when to prune branches: 70-73
   when to prune roots: 136-38
pumice: 164-68
   as a base layer: 144, 160
pyracantha: 106
ramification: 51, 52, 66, 84
repotting
   11 reasons for: 129-35
   calendar: 136-38
   how to: 142-45
   pot refresh: 122
   soil: 162
   the stone trick: 116-17
   tips by tree size: 47-49
   tools: 64
reviving: 154-59
root
   collar: 115-17, 133, 172
   flare: *See* trunk flare
   health: 35, 129-31, 147
   pruning: *See* repotting
   rot: 130-31, 147
sacrifice branch: 78-79, 174
sageretia: 24, 46, 53, 59
saikei: 60
sap: 32-34
   and root pruning: 136-37, 151
   bleeding: 72, 148, 173-74
   flow: 72, 76, 169
satsuki: *See* azalea
science
   fertilizer: 85-97
   of pot size: 119-20
   soil: 161-63
   tree biology: 31-38
seedlings
   growing: 123, 166
   winter care: 29

seeds: *See* propagating
semi-cascade: 59, 118
shari: 174, 198
shohin (one-hander tree): 45-46, 48
shrubs: 37, 69
slanting style: 54
slip potting: 160
soil: 35, 161-68
    components: 164-66
    for tiny bonsai (mame): 46-47
    mix: 163, 167
    recommended by climate: 167-68
    science of: 161-63
    stratification into layers: 167-68
sphagnum moss: 112
    in an air layer: 126-28
spinney: 56
starch
    conversion from sugars: 27, 72
    conversion to sugars: 136
    winter storage: 115, 133, 136
stub: 74-75
subtropical: 29
sugars: 32-34
    and starch: 27, 72, 137
surface roots: 115-17
    and front of tree: 108
    and repotting: 132-34, 143, 144
    in nature: 12, 20
    started by a tourniquet: 170
taper: 20-21, 52-53
    branch taper: 52, 68
    improving trunk taper: 69, 133, 174
    inverse taper: 52, 68, 69
terracotta: 121
tokoname: 121
tools: 62-64
    stainless vs. blackened steel: 63
tourniquet: 169-71
triple-trunk: 55, 145
tropical trees and plants
    climate and location: 26, 28, 179
    defoliation: 82

fertilizing: 100
flowers: 103
pruning: 70
repotting: 137, 151, 153
watering: 179
trunk
    movement: 19, 52-53, 108, 196
    thickening: 16, 83, 120, 172-73
    thickness: 172
    trunk taper: *See* taper
trunk chop: 172-75
    carving: 198
    chopping: 63, 173-74
    developing bonsai: 53, 83, 175
trunk flare: 115-16
    developing: 116, 133, 170
twin-trunk: 55
vermiculite: 164, 166
    for seedlings: 123
vigour: *See* energy
watering: 176-83
    automatic system: 44, 183-88
    box: 182-83
    by climate and season: 179
    can: 180
    guidelines: 177-79
    hose: 181
windswept: 59
wiring
    branches: 189-96
    copper or aluminium: 195
    guy: 193-95
    spiral: 189-92, 195
    tools: 64
    trunk: 196-97
wound sealant: *See* cut paste
xylem: 33-34
    and lignification: 124, 189
    and plant recovery: 155, 158-59
yamadori
    collecting: 61, 198
    style: 61, 197

Acknowledgements and further reading

First and foremost is my infinite gratitude to my wonderful wife for tolerating my long hours on the patio and bonsai mess for the last two decades. And to my three fantastic kids for your enthusiastic participation in bonsai work and YouTube. Thanks also to my sister in law for the gift of my first bonsai twenty years ago.

Thanks infinitely to my truly great parents for your feedback and encouragement, and ultimately for having shaped me into the informal upright that I am today.

Thanks to you: YouTube subscribers for all your support and inciteful comments.

Thanks to these organizations for allowing me to photograph trees in their beautiful bonsai collections:
- Royal Botanical Garden, Madrid
- David Benavente Bonsai Studio

Thanks to these fantastic artists for all their magnificent inspiration:
- Nigel Saunders—The Bonsai Zone
- Peter Chan—Herons Bonsai
- Bjorn Bjorholm—Eisei-en
- Ryan Neil—Bonsai Mirai

References:
- Chan, Peter. (1987). Bonsai Masterclass
- Chan, Peter. (1999). Bonsai: The Art of Growing and Keeping Miniature Trees (original work published 1985)
- De Groot, David. (2016). Principles of Bonsai Design
- Harrington, Harry. (2011). Bonsai Inspirations 1
- Harrington, Harry. (2012). Bonsai Inspirations 2
- Lewis, Colin. (2002). The Art of Bonsai Design
- Matthews, Graeme. (2006). Trees of the World
- Morton, Larry. (2016). Modern Bonsai Practice: 501 Principles of Good Bonsai Horticulture
- Naka, John. (2006). Bonsai Techniques 1 (original work published 1973).
- Norman, Ken. (Spanish edition, 2020). Enciclopedia Visual Del Bonsái (original work published 2013).
- Tomlinson, Harry. (1993). The Complete Book of Bonsai
- Warren, Peter. (2014). Bonsai
- Wohlleben, Peter. (2017). The Hidden Life of Trees

About the author

Dave Seymour has enjoyed the sun-filled culture of Spain for nearly half his life. Originally from the UK, Dave traded in the rainy skies for the scenic beauty of the northern mountains of Madrid. Here, he resides with his wife, three children, and a small forest of potted trees.

With a lifelong career in technical documentation, Dave has developed a knack for making complex information simple and appealing. Combined with his passion for bonsai cultivation, tree photography and videography, this naturally led him to share his expertise with a wider audience. As a YouTube educator and now an author, Dave brings the beautiful nature of bonsai horticulture to enthusiasts around the world.

Discover Dave's captivating bonsai videos on YouTube by searching for "Blue Sky Bonsai", or simply scan this QR code with your phone's camera:

Made in the USA
Las Vegas, NV
01 November 2024